Time and Eternity deals with diff
into relation with traditional theo
and it is marvelous that now the p
losophy of science to the great ben

—John R. Lucas
Fellow of Merton College, Oxford University

William Lane Craig is one the leading philosophers of religion and one of the leading philosophers of time. In this book, he combines his expertise in these areas to produce an original, erudite, and accessible theory of time and God that will be of great interest to both the general public and scholars. It is a rewarding experience to read through this brilliant and well-researched book by one of the most learned and creative thinkers of our era.

—Quentin Smith
Professor of Philosophy, Western Michigan University

Time and Eternity offers a comprehensive discussion of the problems in the concepts of time and eternity on the basis of an extraordinary familiarity with a vast number of recent contributions to this issue from scientists and philosophers. The argument is subtle and precise. Particularly important are the sections on the impact of the different versions of relativity theory on the concept of time.... The book offers a plausible argument for a realistic conception of temporal process and for God's involvement in the temporal distinctions and processes because of his presence in his creation.

—Wolfhart Pannenberg
Professor of Systematic Theology
Ludwig-Maximilians-Universität, München, Germany

In *Time and Eternity*, William Lane Craig defends the remarkable conclusion that "God is timeless without creation and temporal since creation." Craig argues his case philosophically by carefully weighing evidence for and against divine temporality and personhood in light of dynamic versus static theories of time, and this warrants, in turn, a Lorentzian interpretation of special relativity and an objective, mind-independent theory of becoming, including fascinating excursions into Big Bang cosmology and the philosophy of mathematics. As the latest in his series of ground-breaking books, *Time and Eternity* summarizes and extends Craig's previous technical arguments and conveys them to a more general audience. It is a "must-read" for anyone seriously interested in the problem of time and eternity in Christian philosophy.

—Robert Russell
Professor of Theology and Science
Center for Theology and the Natural Sciences
Graduate Theological Union, Berkeley, Calif.

The nature of time is a continuing source of puzzlement both to science and in everyday life. It is also an important issue in theological understandings of the nature of God. In this interesting book, Craig tackles this complex set of topics in a clear way. His discussion of the interrelated scientific, philosophical, and theological issues clears up many previous misconceptions and proposes a plausible understanding of the relation of God to time and eternity that many will find helpful.

—GEORGE ELLIS
Professor of Applied Mathematics
University of Capetown

As a scientist doing theoretical research in gravitational physics and quantum cosmology, I found Dr. Craig's thoughtful book, *Time and Eternity,* highly interesting.

Craig has carefully given arguments defending several different viewpoints for each of the many issues about time that he discusses, followed by critiques in which he emphasizes his own opinion. Reading *Time and Eternity* has forced me to try to develop better arguments for my own opinions (which differ considerably from Craig's), though I do not think that we yet know enough about the subject to settle the issue definitively to everyone's satisfaction.

I am certain that *Time and Eternity* will also stimulate your thinking about this fascinating subject and your appreciation for the God who created time as part of the marvelous universe He has given us.

—DON N. PAGE
Professor of Physics and Fellow of the Cosmology and
Gravitation Program of the Canadian Institute
for Advanced Research
University of Alberta, Edmonton, Alberta, Canada

TIME AND ETERNITY

Exploring God's Relationship to Time

WILLIAM LANE CRAIG

CROSSWAY BOOKS • WHEATON, ILLINOIS
A DIVISION OF GOOD NEWS PUBLISHERS

Time and Eternity

Copyright © 2001 by William Lane Craig

Published by Crossway Books
      a division of Good News Publishers
      1300 Crescent Street
      Wheaton, Illinois 60187

All rights reserved. No part of this publication may be reproduced, stored in a retrieval system, or transmitted in any form by any means, electronic, mechanical, photocopy, recording, or otherwise, without the prior permission of the publisher, except as provided by USA copyright law.

Unless otherwise noted, Scripture references are from the Revised Standard Version. Copyright © 1946, 1953, 1971, 1973 by the Division of Christian Education of the National Council of the Churches of Christ in the U.S.A.

The Scripture reference marked NASB is from the New American Standard Bible® Copyright © The Lockman Foundation 1960, 1962, 1963, 1968, 1971, 1972, 1973, 1975, 1977, 1995. Used by permission.

The Scripture reference marked KJV is from the King James Version.

Cover design: David LaPlaca

Cover photos: PhotoDisc™

Inside photos: Courtesy of the Archives, California Institute of Technology

First printing 2001

Printed in the United States of America

Library of Congress Cataloging-in-Publication Data
Craig, William Lane, 1949 –
    Time and eternity : exploring God's relationship to time / William Lane Craig.
      p. cm.
    Includes bibliographical references and index.
    ISBN 1-58134-241-1 (alk. paper)
    1. God—Immutability. 2. Eternity. 3. Time—Religious aspects—Christianity. I. Title.
BT153.147 C73 2001
231'.4—dc21   00-011716
                                                                          CIP

15	14	13	12	11	10	09	08	07	06	05	04
15	14	13	12	11	10	9	8	7	6	5	4	3	2

For
J. P. MORELAND

Colleague and Friend

*"a mighty man of valor . . .
and the* LORD *is with him"*
(1 Sam. 16:18)

TIME, like an ever-rolling stream,
Bears all its sons away;
They fly, forgotten, as a dream
Dies at the op'ning day.

O GOD, our help in ages past,
Our hope for years to come,
Be Thou our guard while life shall last,
And our eternal home.

— *Isaac Watts*

Contents

Preface   11

1. **TWO VIEWS OF DIVINE ETERNITY**   13
   - I. The Nature of Time
   - II. The Biblical Data on Divine Eternity
   - III. The Importance of Articulating a Theory of Divine Eternity

2. **DIVINE TIMELESSNESS**   29
   - I. Divine Simplicity and Immutability
   - II. Relativity Theory
   - III. The Incompleteness of Temporal Life

3. **DIVINE TEMPORALITY**   77
   - I. The Impossibility of Atemporal Personhood
   - II. Divine Relations with the World
   - III. Divine Knowledge of Tensed Facts

4. **THE DYNAMIC CONCEPTION OF TIME**   115
   - I. Arguments for a Dynamic Conception
     1. The Ineliminability of Tense
     2. Our Experience of Tense
   - II. Arguments against a Dynamic Conception
     1. McTaggart's Paradox
     2. The Myth of Passage

5. **THE STATIC CONCEPTION OF TIME**   167
   - I. Arguments for a Static Conception
     1. Relativity Theory
     2. The Mind-Dependence of Becoming
   - II. Arguments against a Static Conception
     1. "Spatializing" Time
     2. The Illusion of Becoming
     3. The Problem of Intrinsic Change
     4. *Creatio ex Nihilo*

6. **GOD, TIME, AND CREATION**   217
    I. Did Time Begin?
        1. Arguments for the Infinitude of the Past
        2. Arguments for the Finitude of the Past
    II. God and the Beginning of Time
        1. Amorphous Time
        2. Timelessness without Creation

7. **CONCLUSION**   239

APPENDIX: Divine Eternity and God's Knowledge of the Future   243

General Index   266

Scripture and Extra-biblical Literature Index   271

Preface

THE FRENCH HAVE a striking name for God, which, in the French Bible, often stands in the place of our English word "LORD": *l'Eternel*—the Eternal, or the Eternal One. For example, Psalm 106:48 reads,

> Blessed be the Eternal One, the God of Israel,
> From eternity to eternity!
> Let all the people say, "Amen!"
> Praise the Eternal One!

For French-speaking Christians the name *l'Eternel* serves as a constant reminder of the centrality of the divine attribute of eternity. It has become the very name of God.

The present book is written for Christians who want to grapple seriously with the concept of God's eternity. Unlike some other writers on the attributes of God, I am convinced that the best tool we have for really understanding what is meant by the affirmation that God is eternal is not poetry or piety, but analytic philosophy.

Some readers of my study of divine omniscience, *The Only Wise God,* expressed surprise at my remark that someone desiring to learn more about God's attribute of omniscience would be better advised to read the works of Christian philosophers than of Christian theologians.[1] Not only was that remark true, but the same holds for divine eternity. In the Middle Ages students were not allowed to study theology until they had mastered all the other disciplines at the university, but unfortunately today's theologians generally have next to no training in philosophy and science and so are ill-equipped to address in a substantive way the complex issues raised by God's eternity.

As we shall see, divine eternity probably cannot be properly understood without an exploration of the nature of time itself—a daunting prospect! For apart from the idea of God, I know of no concept so profound and so baffling as that of time. To attempt an integration of these two concepts therefore stretches our minds to the very limits of our understanding. But such an exercise will be healthy for us, making us more thoughtful people and deepening our awe and worship of God, the Eternal One.

I have tried to avoid specialist jargon and to define clearly concepts apt

---

[1] William Lane Craig, *The Only Wise God* (Grand Rapids, Mich.: Baker, 1987; rep. ed.: Eugene, Ore.: Wipf & Stock, 2000), 11.

to be unfamiliar to most readers. Nevertheless, I harbor no illusion that this book will be accessible to any interested reader. In writing *The Only Wise God* I found that some concepts are just so difficult that the attempt to simplify can only go so far and that some things will always remain hard to understand. For example, try as one might, it is just impossible to make the Special Theory of Relativity, so central to discussions about time, easy to grasp. But I have tried to state the issues as clearly and simply as I can without sacrificing accuracy.

The present work is a popularization of four scholarly works which are themselves the product of over a dozen years of study of the problem of God and time. An eminent philosopher has remarked that "the problem of time" is virtually unrivaled in "the extent to which it inexorably brings into play all the major concerns of philosophy."[2] Combine the problem of time with "the problem of God," as the study of divine eternity requires, and you have a subject matter which would exhaust a lifetime of study. Readers who are interested in exploring more deeply the nature of time may consult my companion volumes *The Tensed Theory of Time: A Critical Examination* and *The Tenseless Theory of Time: A Critical Examination,* both part of the Synthèse Library series published by Kluwer Academic Publishers of the Netherlands. Those who want a deeper exploration of Relativity Theory from a theistic perspective may want to look at my *Time and the Metaphysics of Relativity,* also available from Kluwer. Finally, my fullest exposition of divine eternity in light of the conclusions of these other works may be found in *God, Time, and Eternity,* published as well by Kluwer.

I am grateful to God for the opportunity, available to so few, to have invested so much study in the effort to sort out divine eternity. And I am grateful to my wife, Jan, for her unflagging support and practical assistance in the execution of this project.

William Lane Craig
Atlanta, Georgia

---

[2] Wilfrid Sellars, "Time and the World Order," *Minnesota Studies in the Philosophy of Science* 8 (1962): 527.

1

TWO VIEWS OF DIVINE ETERNITY

I. The Nature of Time

Time, it has been said, is what keeps everything from happening at once.[1] When you think about it, this definition is probably as good as any other. For it is notoriously difficult to provide any analysis of time that is not in the end circular. If we say, for example, that time is duration, then we shall want to know what duration is. And duration turns out to be some interval of time. So time is some interval of time—not very enlightening! Or if we say that time is a dimension of the world, the points or inhabitants of which are ordered by the relations *earlier than* and *later than*, we may ask for an analysis of those relations so as to distinguish them, for example, from similar relations such as *behind* and *in front of* or *less than* and *greater than*, only to discover that *earlier* and *later*, on pain of circularity, are usually taken to be primitive, or unanalyzable, terms. Perhaps we may define *earlier* and *later* in terms of the notions *past, present,* and *future;* but then this triad is irreducibly temporal in character. Even if we succeed in defining *past* and *future* in relation to the *present,* what is the present except for the time that exists (where "exists" is in the present tense)?

Still, it is hardly surprising that time cannot be analyzed in terms of nontemporal concepts, and the proffered analyses are not without merit, for they do serve to highlight some of time's essential features. For example, most philosophers of time would agree that the *earlier than/later than* relations are essential to time. It is true that in certain high-level theories of physics one sometimes speaks of "imaginary time" or "quantum physical time," which are not ordered by these relations; but it would be far less misleading simply to

---

[1] I first saw this definition in a joke book. But I later discovered that the eminent physicist John Wheeler, in a personal letter to the Russian cosmologist Igor Novikov, had proposed precisely the same definition as his studied analysis of what time is! (Igor D. Novikov, *The River of Time* [Cambridge: Cambridge University Press, 1998], 199).

deny that the geometrical structures posited by the relevant theories really are time at all. Some philosophers of time who deny that the past and future are real or existent have also denied that events or things are related to one another as *earlier than* or *later than;* but such thinkers do affirm the reality of the present as an irreducible feature of time. These features of time are common to our experience as temporal beings, even if ultimately unanalyzable.

Time, then, however mysterious, remains "the familiar stranger."[2] This is the import of St. Augustine's famous disclaimer, "What, then, is time? If no one asks me, I know; but if I wish to explain it to one who asks, I know not."[3]

II. The Biblical Data on Divine Eternity

The question before us concerns the relationship of God to time. The Bible teaches clearly that God is eternal. Isaiah proclaims God as "the high and lofty One who inhabits eternity" (Isa. 57:15). In contrast to the pagan deities of Israel's neighbors, the Lord never came into existence nor will He ever cease to exist. As the Creator of the universe, He was there in the beginning, and He will be there at the end. "I, the LORD, the first, and with the last; I am He" (Isa. 41:4). The New Testament writer to the Hebrews magnificently summarized the Old Testament teaching on God's eternity:

> "Thou, Lord, didst found the earth in the beginning,
> and the heavens are the work of thy hands;
> they will perish, but thou remainest;
> they will all grow old like a garment,
> like a mantle thou wilt roll them up,
> and they will be changed.
> But thou art the same,
> and thy years will never end" (Heb. 1:10-12).

Minimally, then, it may be said that God's being eternal means that God exists without beginning or end. He never comes into or goes out of existence; rather His existence is permanent.[4] Such a minimalist account of divine eternity is uncontroversial.

---

[2] An expression employed by J. T. Fraser, *Time: The Familiar Stranger* (Amherst: University of Massachusetts Press, 1987).
[3] Augustine, *Confessions* 11.14.
[4] For an analysis of what it means to be permanent, see Brian Leftow, *Time and Eternity,* Cornell Studies in the Philosophy of Religion (Ithaca, N.Y.: Cornell University Press, 1991), 133; cf. Quentin Smith, "A New Typology of Temporal and Atemporal Permanence," *Noûs* 23 (1989): 307-330. According to Leftow, an entity is permanent if and only if it exists and has no first or last finite period of existence, and there are no moments before or after it exists.

But there the agreement ends. For the question is *the nature* of divine eternity. Specifically, is God temporal or timeless? God is temporal if and only if He exists in time, that is to say, if and only if His life has phases which are related to each other as earlier and later. In that case, God, as a personal being, has experientially a past, a present, and a future. Given His permanent, beginningless and endless existence, God must be omnitemporal; that is to say, He exists at every moment of time there ever is. I do not mean that He exists at every time at once, which is an incoherent assertion. I mean that if God is omnitemporal, He existed at every past moment, He exists at the present moment, and He will exist at every future moment. No matter what moment in time you pick, the assertion "God exists now" would be literally true at that time.

By contrast, God is timeless if and only if He is not temporal. This definition makes it evident that temporality and timelessness are contradictories: An entity must exist one way or the other and cannot exist both ways at once. Often laymen, anxious to affirm both God's transcendence (His existing beyond the world) and His immanence (His presence in the world), assert that God is both timeless and temporal. But in the absence of some sort of model or explanation of how this can be the case, this assertion is flatly self-contradictory and so cannot be true. If, then, God exists timelessly, He does not exist at any moment of time. He transcends time; that is to say, He exists but He does not exist in time. He has no past, present, and future. At any moment in time at which we exist, we may truly assert that "God exists" in the timeless sense of existence, but not that "God exists now."

Now the question is, does the biblical teaching on divine eternity favor either one of these views? The question turns out to be surprisingly difficult to answer. On the one hand, it is indisputable that the biblical writers typically portray God as engaged in temporal activities, including foreknowing the future and remembering the past; and when they speak directly of God's eternal existence they do so in terms of beginningless and endless temporal duration: "Before the mountains were brought forth, or ever thou hadst formed the earth and the world, from everlasting to everlasting thou art God" (Ps. 90:2). "'Holy, holy, holy is the LORD God Almighty, who was and is and is to come!'" (Rev. 4:8b). After surveying the biblical data on divine eternity, Alan Padgett concludes, "The Bible knows nothing of a timeless divine eternity in the traditional sense."[5]

Defenders of divine timelessness might suggest that the biblical authors lacked the conceptual categories for enunciating a doctrine of divine time-

---

[5] Alan G. Padgett, *God, Eternity, and the Nature of Time* (New York: St. Martin's, 1992), 33.

lessness, so that their temporal descriptions of God need not be taken literally. But Padgett cites the first-century extra-biblical work 2 Enoch 65:6-7 as evidence that the conception of timeless existence was not beyond the reach of biblical writers:

> And then the whole creation, visible and invisible, which the Lord has created, shall come to an end, then each person will go to the Lord's great judgment. And then all time will perish, and afterward there will be neither years nor months nor days nor hours. They will be dissipated, and after that they will not be reckoned (2 Enoch 65:6-7).

Such a passage gives us reason to think that the biblical authors, had they wished to, could have formulated a doctrine of divine timelessness.

Paul Helm raises a more subtle objection to the inference that the authors of Scripture, in describing God in temporal terms, intended to teach that God is temporal.[6] He claims that the biblical writers lacked the "reflective context" for formulating a doctrine of divine eternity. That is to say, the issue (like the issue of geocentrism, for instance) had either never come up for explicit consideration or else simply fell outside their interests. Consider the parallel case of God's relationship to space: Just as the biblical writers describe God in temporal terms, so they describe Him in spatial terms as well:

> "Am I a God at hand, says the LORD, and not a God afar off? Can a man hide himself in secret places so that I cannot see him? says the LORD. Do I not fill heaven and earth? says the LORD" (Jer. 23:23-24).

> Whither shall I go from thy Spirit?
>   Or whither shall I flee from thy presence?
> If I ascend into heaven, thou art there!
>   If I make my bed in Sheol, thou art there!
> If I take the wings of the morning
>   and dwell in the uttermost parts of the sea,
> even there thy hand shall lead me,
>   and thy right hand shall hold me (Ps. 139:7-10).

God is described as existing everywhere in space. Yet most theologians would not take Scripture to teach that God is literally a spatial being. The authors of Scripture were not concerned to craft a metaphysical doctrine of God's relation to space; and parity would require us to say the same of time

---

[6] Paul Helm, *Eternal God* (Oxford: Clarendon, 1988), 5-11.

as well. Padgett considers Helm's point to be well-taken: "The Biblical authors were not interested in philosophical speculation about eternity, and thus the intellectual context for discussing this matter may simply not have existed at that time."[7] Thus, the biblical descriptions of God as temporal may not be determinative for a doctrine of divine eternity.

Moreover, it must be said that the biblical data are not so wholly one-sided as Padgett would have us believe. Johannes Schmidt, whose *Ewigkeitsbegriff im alten Testament* Padgett calls "the longest and most thorough book on the concept of eternity in the OT,"[8] argues for a biblical doctrine of divine timelessness on the basis of creation texts such as Genesis 1:1 and Proverbs 8:22-23.[9] Padgett brushes aside Schmidt's contention with the comment, "Neither of these texts teaches or implies that time began with creation, or indeed say [sic] anything about time or eternity."[10] This summary dismissal is all too quick. Genesis 1:1, which is neither a subordinate clause nor a summary title,[11] states, "In the beginning God created the heavens and the earth." According to James Barr, this absolute beginning, taken in conjunction with the expression, "And there was evening and there was morning, one day" (v. 5), indicating the first day, may very well be intended to teach that the beginning was not simply the beginning of the physical universe but the beginning of time itself, and that, consequently, God may be thought of as timeless.[12] This conclusion is rendered all the more plausible when the Genesis account of creation is read against the backdrop of ancient Egyptian cosmogony.[13] Egyptian cosmogony includes the idea that creation took place at "the first time" (*sp tpy*). John Currid takes both the Egyptian and the Hebrew cosmogonies to involve the notion that the moment of creation is the beginning of time.[14]

Certain New Testament authors may be taken to construe Genesis 1:1 as referring to the beginning of time. The most striking New Testament reflection on Genesis 1:1 is, of course, John 1:1-3: "In the beginning was the Word, and the Word was with God, and the Word was God. He was in the begin-

---

[7] Padgett, *God, Eternity, and the Nature of Time*, 36.
[8] Ibid., 24.
[9] Johannes Schmidt, *Der Ewigkeitsbegriff im alten Testament*, Alttestamentliche Abhandlungen 13/5 (Münster in Westfalen: Verlag des Aschendorffschen Verlagsbuchhandlung, 1940), 31-32.
[10] Padgett, *God, Eternity, and the Nature of Time*, 25.
[11] See exegesis by Claus Westermann, *Genesis 1–11*, trans. John Scullion (Minneapolis: Augsburg, 1984), 97; John Sailhamer, *Genesis*, Expositor's Bible Commentary 2 (Grand Rapids, Mich.: Zondervan, 1990), 21-22.
[12] James Barr, *Biblical Words for Time* (London: SCM Press, 1962), 145-147.
[13] See John D. Currid, "An Examination of the Egyptian Background of the Genesis Cosmogony," *Biblische Zeitschrift* 35 (1991): 18-40.
[14] Ibid., 30.

ning with God; all things were made through him, and without him was not anything made that was made." Here the uncreated Word (*logos*), the source of all created things, was already with God and was God at the moment of creation. It is not hard to interpret this passage in terms of the Word's timeless unity with God—nor would it be anachronistic to do so, given the first-century Jewish philosopher Philo's doctrine of the divine *Logos* (Word) and Philo's holding that time begins with creation.[15]

As for Proverbs 8:22-23, this passage is certainly capable of being read in terms of a beginning of time. The doctrine of creation was a centerpiece of Jewish wisdom literature and aimed to show God's sovereignty over everything. Here Wisdom, personified as a woman, speaks:

> "The LORD possessed me at the beginning of His way,
> Before His works of old.
> From everlasting I was established,
> From the beginning, from the earliest times of the earth" (NASB).

The passage, which doubtless looks back to Genesis 1:1, is brimming with temporal expressions for a beginning. R. N. Whybray comments,

> It should be noted how the writer . . . was so insistent on pressing home the fact of Wisdom's unimaginable antiquity that he piled up every available synonym in a deluge of tautologies: *rēs'šît*, **beginning**, *qedem*, **the first**, *mē'āz*, **of old**, *mē 'olām*, **ages ago**, *mērō'š*, **at the first** or "from the beginning" (compare Isa. 40.21; 41.4, 26), *miqqadᵉ mê'āreṣ*, **before the beginning of the earth**: the emphasis is not so much on the *mode* of Wisdom's coming into existence, . . . but on the *fact* of her antiquity.[16]

The expressions emphasize, however, not Wisdom's mere antiquity, but that there was a beginning, a departure point, at or before which Wisdom existed. This was a departure point not merely for the earth but for time and the ages; it was simply the beginning. Plöger comments that through God's creative work "the possibility of speaking of 'time' was first given; thus, before this

---

[15] On the beginning of time with creation, see Philo of Alexandria, *On the Creation of the Cosmos according to Moses*, trans. with an introduction and commentary by David T. Runia, Philo of Alexandria Commentary Series 1 (Leiden: E. J. Brill, forthcoming); cf. Richard Sorabji, *Time, Creation and the Continuum* (Ithaca, N.Y.: Cornell University Press, 1983), 203-209. For a discussion of the similarities between John's prologue and Philo's *De opificio* 16-19, in which his *logos* doctrine of creation is described, see C. H. Dodd, *The Interpretation of the Fourth Gospel* (Cambridge: Cambridge University Press, 1953), 66-73, 276-277.

[16] R. N. Whybray, *Proverbs*, New Century Bible Commentary (Grand Rapids, Mich.: Eerdmans, 1994), 131-132.

time, right at the beginning, Wisdom came into existence through Yahweh [the LORD]."[17] The passage was so understood by other ancient writers. The Septuagint Greek translation of the Old Testament renders *mē 'ōlām* in Proverbs 8:23 as *pro tou aiōnios* (before time), and Sirach 24:9 has Wisdom say, "Before the ages, in the beginning, he created me, and for all ages I shall not cease to be" (cf. 16:26; 23:20).

Significantly, certain New Testament passages also seem to affirm a beginning of time. This would imply just the same sort of timelessness "before" the creation of the world which Padgett sees in 2 Enoch "after" the end of the world. For example, we read in Jude 25, "to the only God, our Savior through Jesus Christ our Lord, be glory, majesty, dominion, and authority, *before all time* and *now* and *for ever*" (*pro pantos tou aiōnos kai nun kai eis pantas tous aiōnas*) (emphasis added). The passage contemplates an everlasting future duration but affirms a beginning to past time and implies God's existence, using an almost inevitable *façon de parler*, "before" time began. Similar expressions are found in two intriguing passages in the Pastoral Epistles. In Titus 1:2-3, in a passage laden with temporal language, we read of those chosen by God "in hope of eternal life [*zōēs aiōniou*] which God, who never lies, promised before age-long time [*pro chronon aiōnion*] but manifested at the proper time [*kairois idiois*]" (author's translation). And in 2 Timothy 1:9 we read of God's "purpose and grace, which were given to us in Christ Jesus before age-long time [*pro chronon aiōnion*], but now [*nun*] manifested by the appearing of our Savior Christ Jesus" (author's translation). Arndt and Gingrich render *pro chronon aiōnion* as "before time began."[18] Similarly, in 1 Corinthians 2:7 Paul speaks of a secret, hidden wisdom of God, "which God decreed before the ages [*pro tōn aiōnōn*] for our glorification." Such expressions are in line with the Septuagint, which describes God as "the one who exists before the ages [*ho hyparchōn pro tōn aiōnōn*]" (LXX Ps. 54:20 [Ps 55:19]). Expressions such as *ek tou aiōnos* or *apo tōn aiōnōn* might be taken to mean merely "from ancient times" or "from eternity." But these should not be conflated with *pro* expressions. That such *pro* constructions are to be taken seriously and not merely as idioms connoting "for long ages" (cf. Rom. 16:25: *chronois aiōniois*) is confirmed by the many similar expressions concerning God and His decrees "before the foundation of the world" (*pro*

---

[17] Otto Plöger, *Sprüche Salomos*, Biblisches Kommentar altes Testaments 17 (Neukirchen-Vluyn: Neukirchner Verlag, 1984), 92. Cf. Meinhold's comment: "Its [time's] beginning is set at the first act of creation" (Arndt Meinhold, *Die Sprüche*, vol. 1, Zürcher Bibelkommentare [Zürich: Theologischer Verlag Zürich, 1991], 144).

[18] Walter Bauer, *A Greek-English Lexicon of the New Testament*, trans. and ed. W. F. Arndt and F. W. Gingrich, s.v. "aionios."

kataboles kosmou) (John 17:24; Eph. 1:4; 1 Pet. 1:20; cf. Rev. 13:8). Evidently it was a common understanding of the creation described in Genesis 1:1 that the beginning of the world was coincident with the beginning of time or the ages; but since God did not begin to exist at the moment of creation, it therefore followed that He existed "before" the beginning of time. God, at least "before" creation, must therefore be atemporal.

Thus, although scriptural authors speak of God as temporal and everlasting, there is some evidence, at least, that when God is considered in relation to creation He must be thought of as the transcendent Creator of time and the ages and therefore as existing beyond time. It may well be the case that in the context of the doctrine of creation the biblical writers were led to reflect on God's relationship to time and chose to affirm His transcendence. Still the evidence is not clear, and we seem forced to conclude with Barr that "if such a thing as a Christian doctrine of time has to be developed, the work of discussing it and developing it must belong not to biblical but to philosophical theology."[19]

III. The Importance of Articulating a Theory of Divine Eternity

If the biblical data concerning God's relationship to time are indeterminative, then why, it may be asked, not simply rest with the biblical affirmation of God's beginningless and endless existence, instead of entering the speculative realms of metaphysics in an attempt to articulate a doctrine of God and time? At least two responses may be given to this question. First, *the biblical conception of God has been attacked precisely on the grounds that no coherent doctrine of divine eternity can be formulated.* Two examples come immediately to mind. In his *God and the New Physics,* Paul Davies, a distinguished physicist who was awarded the million-dollar Templeton Prize for Progress in Religion for his many popular books relating science and religion, argues that God, as traditionally understood, can be neither timeless nor temporal. On the one hand, God cannot be timeless because such a being "cannot be a personal God who thinks, converses, feels, plans, and so on for these are all temporal activities."[20] Such a God could not act in time, nor could He be considered a self and, hence, a person. Davies adds, "The difficulty is particularly acute for Christians, who believe that at some specific moment in human history, God became incarnate and set about saving Man."[21] On the other

---

[19] Barr, *Biblical Words for Time,* 149.
[20] Paul Davies, *God and the New Physics* (New York: Simon and Schuster, 1983), 133-134; cf. 38-39.
[21] Unpublished transcript of a lecture courtesy of Paul Davies.

hand, according to Davies, God cannot be a temporal being because He would then be subject to the laws of Relativity Theory governing space and time and so could not be omnipotent; nor could He be the Creator of the universe, since in order to create time and space, God must transcend time and space. Davies insists,

> God the Creator, by his very nature, must transcend space and time. . . . the coming into being of the physical universe involved the coming into being of space and time as well as matter. I can't emphasize this too strongly and so if we wish to have a God who is in some sense responsible for the origin of the universe or for the universe, then this God must lie outside of the space and time which is being created.[22]

The logical conclusion of Davies's dilemma is that God as the Bible portrays Him does not exist. The importance of this dilemma has grown in Davies's thinking over the years; he has recently written, "No attempt to explain the world, either scientifically or theologically, can be considered successful until it accounts for the paradoxical conjunction of the temporal and the atemporal, of being and becoming."[23]

A second example of such an attack on the biblical conception of God is the critique of God as Creator set forth by Stephen Hawking, one of the most celebrated mathematical physicists of the twentieth century, in his runaway best-seller *A Brief History of Time.* Hawking believes that in the context of standard Big Bang cosmology it makes sense to appeal to God as the Creator of the space-time universe, since according to that theory space-time had a beginning point, called the initial singularity, at which the universe originated.[24] By introducing imaginary numbers (multiples of $\sqrt{-1}$) for the time variable in the equations describing the very early universe, Hawking eliminates the singularity by "rounding off," as it were, the beginning of space-time. Instead of having a beginning point akin to the apex of a cone, space-time in its earliest state in Hawking's theory is like the rounded tip of a badminton birdie. Like the surface of a sphere, it has no edge at which you must stop. Hawking is not at all reluctant to draw theological conclusions from his model:

---
[22] Ibid.
[23] Paul Davies, *The Mind of God* (New York: Simon and Schuster, 1992), 38.
[24] Space-time is simply that four-dimensional continuum composed of the three familiar spatial dimensions—length, width, and height—plus the dimension of time.

There would be no singularities at which the laws of science broke down and no edge of space-time at which one would have to appeal to God or some new law to set the boundary conditions for space-time. . . . The universe would be completely self-contained and not affected by anything outside itself. It would be neither created nor destroyed. It would just BE. . . .

The idea that space and time may form a closed surface without boundary . . . has profound implications for the role of God in the affairs of the universe. . . . So long as the universe had a beginning, we could suppose it had a creator. But if the universe is really completely self-contained, having no boundary or edge, it would have neither beginning nor end. What place, then, for a creator?[25]

The success of Hawking's gambit to eliminate the Creator of the universe hinges crucially on the legitimacy of his concept of "imaginary time." Since on Hawking's view imaginary time is indistinguishable from a spatial dimension, devoid of temporal becoming and *earlier than/later than* relations, the four-dimensional space-time world just subsists, and there is nothing for a Creator to do.

Both Davies and Hawking's writings have been enormously influential in popular culture as well as in scientific thinking. An adequate answer to the challenges they pose to biblical theism requires a coherent theory of divine eternity and God's relation to time.

The second reason why it is incumbent upon the philosophical theologian to articulate a doctrine of God and time is that *a great deal of careless writing has already been done on this topic.* The question is not whether orthodox believers will address the issue, but whether they will address it responsibly. It is inevitable that when Christians think about God's eternity or knowledge of the future or of our "going to be with the Lord in eternity," they will form conceptions of how God relates to time. These are usually confused and poorly thought through, a situation often exacerbated by pronouncements from the pulpit concerning divine eternity. Unfortunately, popular authors frequently compound the problem in their treatments of God and time.

Again, two examples will suffice. Philip Yancey is an enormously popular Christian author. In his award-winning book *Disappointment with God,* Yancey attempts to come to grips with the apparently gratuitous evil permitted by God in the world. The centerpiece of his solution to the problem is his understanding of God's relationship to time.[26] Unfortunately,

---

[25] Stephen Hawking, *A Brief History of Time* (New York: Bantam Books, 1988), 136, 140-141.
[26] Philip Yancey, *Disappointment with God* (Grand Rapids, Mich.: Zondervan, 1988), 194-199.

Yancey's view is a self-contradictory combination of two different positions based on a pair of confused analogies. On the one hand, appealing to the Special Theory of Relativity, Yancey wants to affirm that a being coextensive with the universe would know what is happening from the perspective of any spatially limited observer in the universe. But, contrary to Yancey, the fact that local observers have varying perspectives has nothing to do with relativity at all, but rather with the finite velocity of light. Localized observers can only form what cosmologists call a "world picture" of the universe: As they look out into space they are seeing astronomical events, not as they are occurring simultaneously with local events but as they were in the past. Local observers at distant places in the universe will thus have different world pictures. What they cannot form is a "world map," that is, a picture of what is happening in the universe simultaneously with events in their vicinity. A cosmic observer such as Yancey imagines would, however, be able to form a world map precisely because he is not spatially localized. Such a cosmic observer would experience the lapse of worldwide cosmic time and would be able to know what is happening now anywhere in the universe. If we deny him such a cosmic perspective and grant to him only a combination of local perspectives, then he becomes a pitiful schizophrenic, lacking all unity of consciousness and possessing only an infinitely fragmented array of local consciousnesses—hardly an adequate analogy for God! In any case, the salient point is that such a being would be temporal and would experience the flow of time. Such an understanding is inconsistent with Yancey's second analogy of the relation between the time of an author and the time of the characters in his book or film. "We see history like a sequence of still frames, one after the other, as in a motion picture reel; but God sees the entire movie at once, in a flash."[27] The analogy is problematic, since characters in novels and films do not really exist, and so neither do their "times" exist. Hence, there just is no relation between, say, the time of Shakespeare and the time of Hamlet. But again, the salient point is that this analogy points in a direction opposite the first, to an understanding of time as static, like a film lying in the can or a novel sitting on the shelf, with a timeless God existing outside the temporal dimension. Yancey's two analogies thus issue in a self-contradictory view of divine eternity—unless, perhaps, he makes the extravagant move of construing eternity as a sort of hyper-time, a higher, second-order time dimension in which our temporal

---

[27] Ibid., 197.

dimension is embedded—and so provides no adequate solution to the problem of disappointment with God.[28]

Our second example is the popular science writer Hugh Ross, who apparently does make so bold as to affirm that God exists and operates in hyper-time. Explicitly rejecting the Augustinian-Thomistic doctrine of divine timelessness, Ross affirms that "The Creator's capacities include at least two, perhaps more, time dimensions."[29] In attempting to solve the problem of God's creating time (raised by Davies above), Ross asserts that God exists in a sort of hyper-time, in which He created our space-time universe. Unfortunately, Ross does not accurately represent this notion. A divine hyper-time would be a dimension at each of whose moments our entire time dimension exists or not. On a diagram, it would be represented by a line perpendicular to the line representing our dimension (Fig. 1.1):

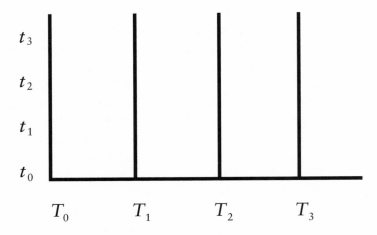

Fig. 1.1: At successive moments of hyper-time T, our entire time series t exists.

But Ross misconstrues the nature of hyper-time, representing God's time on his diagram by a line parallel, rather than perpendicular, to the line representing our temporal dimension.[30] Fig. 1.2 reproduces Ross's Fig. 7.1:

---

[28] For another popular misuse of Relativity Theory in the service of theology, see Anthony Campolo, *A Reasonable Faith* (Waco, Tex.: Word, 1983), 128-134. Campolo hopes to solve problems of predestination and the intermediate state of the dead by appeal to the relativity of simultaneity—as though God were a physical object in an inertial frame moving at the speed of light!
[29] Hugh Ross, *Beyond the Cosmos* (Colorado Springs: NavPress, 1966), 24.
[30] Ibid., 62.

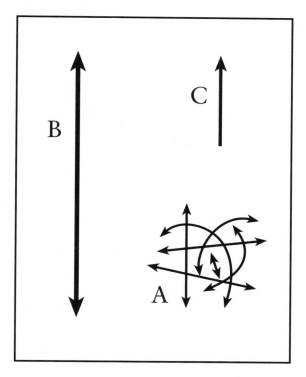

Fig. 1.2: *B* represents God's infinite time line, while *C* represents our finite time line. *A* erroneously depicts other alleged time lines.

What Ross's diagram implies is that God's temporal dimension is actually the same as ours, but that He pre-exists for infinite time prior to the creation of the universe. This is, in fact, a classical, Newtonian view of God and time. Newton believed that God existed from eternity past in absolute time and at some moment created the physical universe. The proper distinction to be drawn on such a view is not between two dimensions of time, but rather, as Newton put it, between absolute time and our relative, physical measures of time. In affirming God's infinite pre-existence, Ross must face the old question that dogged Newtonians: Why would God delay for infinite time the creation of the universe?

In two places Ross suggests that the two dimensions of time may have the geometry of the surface of a hemisphere, our time being represented by the equator and God's time by the longitudinal lines (Fig. 1.3).[31]

---
[31] Ibid., 57, 151.

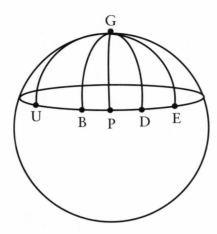

Fig. 1.3: *UE* represents the time dimension of the universe. *G* represents God. *GU, GB,* etc., must then represent separate time lines on which God exists.

Such a daring model is, however, misconceived. For then it is our time which is the hyper-time in which God's temporal dimension is embedded, since there is one line representing our time but many lines for God's. Moreover, it is incorrect to situate God at the pole of the hemisphere, as Ross does, for this would be to treat His time as the embedding hyper-time; in fact, He must exist at all the points on each of His longitudinal time lines. Since these divine time lines endure through successive moments of our hyper-time, they cannot represent lines of divine causal influence, as Ross thinks. Finally, such a view makes our time circular, which contradicts the Judaeo-Christian conception of time. This unwelcome conclusion could be averted only by making our time finite in extent, which contradicts the Christian doctrine of immortality. In short, Ross's views, while ingenious, are neither coherent nor consistent with orthodox theology. What makes this conclusion disturbing is Ross's repeated claim that Christian doctrines such as the Trinity and the incarnation are not logically coherent unless formulated in more than four dimensions. I suspect that, for Ross, talk of God's extra-dimensionality is but a *façon de parler* for God's transcending space and time—but then he has expressed himself in a most misleading way, which is bound to create confusion and still leaves us with no clear understanding of God's relationship to time.

Examples could be multiplied to show the way in which popular expositions of divine eternity have promoted error or confusion. The philosopher Max Black once remarked that "a rough measure of the philosophical importance of a concept is the amount of nonsense written about it. Judged by this test the concept of time comes somewhat ahead of the concept of space and behind the concept of deity."[32] Combine time and deity and you really have something both important and difficult to write about! If we are to move beyond the nonsense, clear, rigorous thinking—not silence—is called for on this issue.

We therefore have good reason to turn to philosophical theology for an articulation of a doctrine of divine eternity. When we do so, as the above discussions remind us, we shall have to keep an eye on science as well as philosophy. Of course, for the Christian, one's theory of divine eternity will be held tentatively, as our best effort to understand how God relates to time, rather than dogmatically, as if it were the teaching of Scripture. Scripture teaches that God exists beginninglessly and endlessly; now it is up to us to figure out what that implies.

Recommended Reading*

Padgett, Alan G. *God, Eternity, and the Nature of Time,* chapter 2. New York: St. Martin's, 1992.

Helm, Paul. *Eternal God,* pp. 9-11. Oxford: Clarendon, 1988.

*Items in "Recommended Reading" lists appear in the order in which their subjects were discussed in the chapter, rather than in alphabetical order.

---

[32] Max Black, review of *The Natural Philosophy of Time,* by G. J. Whitrow, in *Scientific American* 206 (April 1962), 179.

2

DIVINE TIMELESSNESS

"WHATEVER INCLUDES AND possesses the whole fulness of interminable life at once and is such that nothing future is absent from it and nothing past has flowed away, this is rightly judged to be eternal," wrote the medieval theologian Boethius.[1] On such an understanding of divine eternity God transcends time altogether. But what reasons can be given for adopting such an understanding of God's eternity? In the next two chapters we shall examine what I consider to be the most important arguments for divine timelessness and for divine temporality. In this chapter we shall look at what I deem to be the most important arguments on behalf of the view that God is timeless.

I. Divine Simplicity and Immutability

EXPOSITION

Traditionally, Christian theologians such as Thomas Aquinas argued for God's timelessness on the basis of His absolute simplicity and immutability. The argument can be easily formulated. As a first premise, we assume either

    1. God is simple

or

    1'. God is immutable.

Then we add

    2. If God is simple or immutable, then He is not temporal,

---

[1] Boethius, *Consolation of Philosophy* 5. pr. 6. 25-31.

from which we can logically deduce

    3. Therefore, God is not temporal.

Since temporality and timelessness are, as we have seen, contradictories, it follows that

    4. Therefore, God is timeless.

Since this is a logically valid argument, the only question to consider is whether the premises of the argument are true.

Critique

Consider premise (2) above. The doctrine of divine simplicity states that God has absolutely no composition in His nature or being. Thus, the notion of simplicity operative here is the polar opposite of complexity. God is said to be an absolutely undifferentiated unity. This medieval doctrine is not popular among theologians today, and even when Christians do give lip service to it, they usually do not appreciate how truly radical the doctrine is. It implies not merely that God does not have parts, but that He does not possess even distinct attributes. In some mysterious way His omnipotence is His goodness, for example. He stands in no relations whatsoever. Thus, He does not literally love, know, or cause His creatures. He is not really composed of three distinct persons, a claim notoriously difficult to reconcile with the doctrine of the Trinity. His nature or essence is not even distinct from His existence, an assertion which led to the very difficult doctrine that God's essence just *is* existence; He is, Thomas Aquinas tells us, the pure act of existing.

Now if God is simple in the way described, it obviously follows that He cannot be temporal, for a temporal being is related to the various times at which it exists: It exists at t_1 and at t_2, for example. But a simple being stands in no real relations, as we have seen. Moreover, a temporal being has phases of its life which are not identical but rather are related to one another as earlier and later. But an absolutely simple being could not stand in such relations and so must have its life, as Boethius put it, "all at once" (*totum simul*).

Similarly, if God is immutable, then even if He is not simple He still cannot be temporal. Like simplicity, the immutability affirmed by the medieval theologians is a radical concept: utter immobility. God cannot change in *any* respect. He never thinks successive thoughts, He never performs successive actions, He never undergoes even the most trivial alteration. God not only cannot undergo intrinsic change, He cannot even change extrinsically by being related to changing things.² But obviously a temporal being undergoes at least extrinsic change in that it exists at different moments of time and, given the reality of the temporal world, co-exists with different sets of temporal beings as they undergo intrinsic change. Even if we relax the definition of "immutable" to mean "incapable of intrinsic change," or the even weaker concept "intrinsically changeless," an immutable God cannot be temporal. For if God is temporal, He at the very least changes in that He is constantly growing older—not physically, of course, but in the purely temporal sense of constantly adding more years to His life. Moreover, God would be constantly changing in His knowledge, knowing first that "It is now t_1" and later that "It is now t_2." God's foreknowledge and memory must also be steadily changing, as anticipated events transpire and become past. God would constantly be performing new actions, at t_1 causing the events at t_1, and at t_2 causing the events at t_2. Thus, a temporal God cannot be changeless. It follows, then, that if God is immutable, He is timeless.

Thus, God's timelessness can be deduced from either His simplicity or His immutability. Is this a good reason for thinking that God is timeless? That all depends on whether we have any good reason to think that God is simple or immutable. Here we run into severe difficulties. For the doctrines of divine simplicity and immutability are even more controverted than the doctrine of divine eternity. To try to prove divine timelessness via divine simplicity or immutability, therefore, takes on the air of trying to prove the obvious via the less obvious. More specifically, the doctrines of divine simplicity and immutability as explained above find absolutely no support in Scripture, which at most speaks of God's immutability in terms of His faithfulness and unchanging character (Mal. 3:6; Jas. 1:17). Philosophically, there seem to be no good reasons to embrace these radical doctrines, and weighty objections have been lodged against them.³ These need not be discussed here; the point

---

² An intrinsic change is a non-relational change, involving only the subject. For example, an apple changes from green to red. An extrinsic change is a relational change, involving something else in relation to which the subject changes. For example, I become shorter than my son, not by undergoing an intrinsic change in my height, but by being related to him as he undergoes intrinsic change in his height. I change extrinsically from being taller than John to being shorter than John because he is growing.

³ Thomas Aquinas's famous argument for God from contingent beings (beings whose essence is distinct from their existence) leads at best, I think, to a being whose essence is such that it is necessarily instantiated, a metaphysically necessary being. But to say that God does not have distinct properties seems patently false: Omnipotence is not the same property as goodness, for a being may have one and not the other. To respond that these properties differ in our conception only, as manifestations of a single divine property, just as, say, "the morning star" and "the evening star" have different senses but both refer to the same reality (Venus) is inadequate. For *being the morning star* and *being the evening star* are distinct properties both possessed by Venus; in the same way, *being omnipotent* and *being good* are not different senses for the same property (as are, say, *being even* and *being divisible by two*) but are clearly distinct properties. To say that God is His essence seems to make God into a property, which is incompatible with His being a living, concrete being. Moreover, if God is not distinct from His essence, then God

is that premises (1) and (1') above are even less plausible and more difficult to prove than (4), so that they do not constitute good grounds for believing (4). Thus, while we may freely admit that a simple or immutable God must be timeless, we have even less reason to think God simple or immutable than to think Him timeless and so can hardly infer that He is timeless on the basis of those doctrines.

II. Relativity Theory

EXPOSITION

The branch of physics most directly concerned with the analysis of the nature of time and space is Relativity Theory, the brainchild of Albert Einstein. There are two theories of relativity, the restricted or Special Theory of Relativity (STR), which Einstein formulated in 1905, and the General Theory of Relativity (GTR), which he completed in 1915. According to physicist Hermann Bondi, "there is perhaps no other part of physics that has been checked and tested and cross-checked quite as much as the Theory of Relativity."[4] The predictions of both STR and GTR have been verified without fail to a fantastic degree of precision. Any adequate theory of God's relationship to time must therefore take account of what these theories have to say about the nature of time. When we explore what STR has to say about the nature of time and particularly about simultaneity, however, a significant objection to divine temporality arises.

In order to grasp this objection, we need to have some understanding of STR. Although the mathematics of STR are not highly sophisticated, nonetheless the *concepts* of time and space defined by the theory are so strange and counterintuitive that most people, I venture to say, find them nearly inconceivable. Undaunted, I shall attempt to explain in as simple a way as possible what

---

cannot know or do anything different than what He knows and does, in which case everything becomes necessary. To respond that God is perfectly similar in all logically possible worlds which we can imagine but that contingency is real because God stands in no real relations to things is to make the existence or non-existence of creatures in various possible worlds independent of God and utterly mysterious. To say that God's essence just is His existence seems wholly obscure, since then there is in God's case no entity that exists; there is just the existing itself without any subject. For further critique, see Christopher Hughes, *On a Complex Theory of a Simple God*, Cornell Studies in the Philosophy of Religion (Ithaca, N.Y.: Cornell University Press, 1989); Thomas V. Morris, *Anselmian Explorations* (Notre Dame, Ind.: University of Notre Dame Press, 1987), 98-123.

Divine immutability is sometimes said to be a necessary correlate of divine perfection. But this seems clearly incorrect. A perfect being need not change "vertically," so to speak, on the scale of perfection and, hence, for the worse, but could change "horizontally," remaining equally perfect in both states. For example, for God to change from knowing "It is now t_1" to knowing "It is now t_2" is not a change for the worse in God; on the contrary, it is a sign of His perfection that He always knows what time it is. It is also not clear that divine immutability (as opposed to mere changelessness) is compatible with divine freedom—suppose God wanted to change?

[4] Hermann Bondi, *Relativity and Common Sense* (New York: Dover, 1964), 168.

Isaac Newton
"And thus much concerning God; to discourse of whom from the appearances of things, does certainly belong to Natural Philosophy."

Einstein's theory holds with regard to the nature of time and space, so that we may then understand what impact this has on our conception of divine eternity.

Let us begin with a historical retrospect. The physics which prevailed up until the reception of Relativity Theory was Newtonian physics, whose foundations were laid by Isaac Newton, perhaps the greatest scientist of all time, in his epochal *Philosophiae naturalis principia mathematica* (1687). In the *Scholium* to his set of Definitions leading off the *Principia,* Newton explains his concepts of time and space. In order to clarify these concepts, Newton draws a distinction between *absolute* time and space and *relative* time and space:

> I. Absolute . . . time, of itself, and from its own nature, flows equably without relation to anything external, and by another name is called duration: relative . . . time, is some sensible and external (whether accurate or unequable) measure of duration by the means of motion, which is commonly used instead of true time; such as an hour, a day, a month, a year.

> II. Absolute space, in its own nature, without relation to anything external, remains always similar and immovable. Relative space is some movable dimension or measure of the absolute spaces; which our senses determine by its position to bodies; and which is commonly taken for immovable space; such is the dimension of a subterraneous, an aerial, or celestial space, determined by its position in respect of the earth.[5]

---

[5] Isaac Newton, *Sir Isaac Newton's "Mathematical Principles of Natural Philosophy" and his "System of the World,"* trans. Andrew Motte, rev. with an appendix by Florian Cajori, 2 vols. (Los Angeles: University of California Press, 1966), 1:6.

Fundamentally, Newton is here distinguishing between time and space themselves and our *measures* of time and space. Relative time is the time determined or recorded by clocks and calendars of various sorts; relative space is the length or area or volume determined by instruments such as rulers or measuring cups. As Newton says, these relative quantities may be more or less accurate measures of time and space themselves. Time and space themselves are absolute in the sense that they just are the quantities themselves which we are trying to measure with our physical instruments.

There is, however, another sense in which Newton held time and space to be absolute. They are absolute in the sense that they are unique. There is one, universal time in which all events come to pass with determinate duration and in a determinate sequence, and one, universal space in which all physical objects exist with determinate shapes and in a determinate arrangement. Thus Newton says that absolute time "of itself, and from its own nature, flows equably without relation to anything external," and absolute space "in its own nature, without relation to anything external, remains always similar and immovable." Relative times and spaces are many and variable, but not time and space themselves.

On the basis of his definitions of time and space, Newton went on to define absolute versus relative place and motion:

> III. Place is a part of space which a body takes up, and is according to the space, either absolute or relative. . . .
>
> IV. Absolute motion is the translation of a body from one absolute place into another; and relative motion, the translation from one relative place into another.[6]

By "translation" Newton means "transporting" or "displacement." Absolute place is the volume of absolute space occupied by an object, and absolute motion is the displacement of a body from one absolute place to another. An object can be at relative rest and yet in absolute motion. Newton gives the example of a piece of a ship, say, the mast. If the mast is firmly fixed, then it is at rest relative to the ship; but the mast is in absolute motion if the ship is moving in absolute space as it sails along. Thus, two objects can be at rest relative to each other, but both moving in tandem through absolute space (and thus moving absolutely). Similarly, two objects—say, two asteroids—

---

[6] Ibid., 1:6-7.

could be in motion relative to each other and yet one of them at rest in absolute space.

In Newtonian physics there is already a sort of relativity. A body which is in uniform motion (that is, no accelerations or decelerations occur) serves to define an inertial frame, which is just a relative space in which a body at rest remains at rest and a body in motion remains in motion with the same speed and direction. Newton's ship sailing uniformly along would thus define an inertial frame. Although Newton postulated the existence of an absolute inertial frame, namely, the reference frame of absolute space, nevertheless it was impossible for observers in inertial frames which were moving in absolute space to determine experimentally that they were in fact moving. If someone's relative space were moving uniformly through absolute space, that person could not tell whether he was at absolute rest or in absolute motion. By the same token, if his relative space were at rest in absolute space, he could not know that he was at absolute rest rather than in absolute motion. He could know that his inertial frame was in motion relative to some other observer's inertial frame (say, another passing ship), but he could not know if either of them were at absolute rest or in absolute motion. Thus, within Newtonian physics an observer could measure only the relative motion of his inertial system, not its absolute motion.

This sort of relativity was known long before Newton. Galileo, for example, understood it and provided a delightful illustration of it:

> For a final indication of the nullity of the experiments brought forth, this seems to me the place to show you a way to test them all very easily. Shut yourself up with some friend in the main cabin below decks on some large ship, and have with you there some flies, butterflies, and other small flying animals. Have a large bowl of water with some fish in it; hang up a bottle that empties drop by drop into a wide vessel beneath it. With the ship standing still, observe carefully how the little animals fly with equal speed to all sides of the cabin. The fish swim indifferently in all directions; the drops fall into the vessel beneath; and, in throwing something to your friend, you need throw it no more strongly in one direction than another, the distances being equal; jumping with your feet together, you pass equal spaces in every direction. When you have observed all these things carefully (though there is no doubt that when the ship is standing still everything must happen in this way), have the ship proceed with any speed you like, so long as the motion is uniform and not fluctuating this way and that. You will discover not the least of change in all the effects named, nor could you tell from any of them whether the ship was moving or standing still. In jumping, you will pass on

the floor the same spaces as before, nor will you make larger jumps toward the stern than toward the prow, even though the ship is moving quite rapidly, despite the fact that during the time that you are in the air the floor under you will be going in a direction opposite to your jump. In throwing something to your companion, you will need no more force to get it to him whether he is in the direction of the bow or the stern, with yourself situated opposite. The droplets will fall as before into the vessel beneath without dropping toward the stern, although while the drops are in the air the ship runs many spans. The fish in their water will swim toward the front of their bowl with no more effort than toward the back, and will go with equal ease to bait placed anywhere around the edges of the bowl. Finally the butterflies and flies will continue their flights indifferently toward every side, nor will it ever happen that they are concentrated toward the stern, as if tired out from keeping up with the course of the ship, from which they will have been separated during long intervals by keeping themselves in the air. And if smoke is made by burning some incense, it will be seen going up in the form of a little cloud, remaining still and moving no more toward one side than the other. The cause of all these correspondences of effects is the fact that the ship's motion is common to all the things contained in it, and to the air also.[7]

In this case, so long as the ship continues in uniform motion, the relative space occupied by the ship's cabin defines an inertial frame which may or may not be at absolute rest and relative to which the butterflies and fish and smoke move as though it were at absolute rest. There is no way to tell. In honor of Galileo, this sort of relativity is usually called Galilean Relativity.

Although Galilean Relativity was enunciated more than 400 years ago, most laymen still have not absorbed it (much to the dismay of science teachers!). People still puzzle over whether they could save themselves from being smashed to death in a freely falling elevator by leaping into the air just before it hits the ground—forgetting that even if they reverse their motion relative to the inertial frame of the elevator, they are still plunging downward relative to the inertial frame of the ground!

Newtonian physics prevailed all the way up through the end of the nineteenth century. The two great domains of nineteenth-century classical physics were Newton's mechanics (the study of the motion of bodies) and James Clerk Maxwell's electrodynamics (the study of electro-magnetic radiation, including light). The quest of physics at the end of the nineteenth century was to formulate mutually consistent theories of these two domains. The prob-

---

[7] Galileo Galilei, *Dialogue Concerning the Two Chief World Systems—Ptolemaic and Copernican.*, trans. Stillman Drake (Berkeley: University of California Press, 1962), 186-188.

lem was that although Newton's mechanics was characterized, as we have seen, by relativity, Maxwell's electrodynamics was not. It was widely held that light (and other forms of electro-magnetic radiation) consisted of waves, and, since waves had to be waves of something (for example, sound waves are waves of the air; ocean waves are waves of the water), light waves had to be waves of an invisible, all-permeating substance dubbed "the aether." As the nineteenth century wore on, the aether was divested of more and more of its properties until it became virtually characterless, serving only as the medium for the propagation of light. Since the speed of light had been measured and since light consisted of waves in the aether, the speed of light was absolute; that is to say, unlike moving bodies, light's velocity was determinable relative to an absolute frame of reference, the aether frame. To be sure, in the Newtonian scheme of things, moving bodies *possessed* absolute velocities relative to this frame, but within an inertial frame there was no way to *measure* what it was. By contrast, since waves move through their medium at a constant speed regardless of how fast the object which caused them is moving, light had a determinable, fixed velocity. So electrodynamics, unlike mechanics, was not characterized by relativity.

But now it seemed that one could use electrodynamics to eliminate Galilean Relativity. Since light moved at a fixed rate through the aether, one could, by measuring the speed of light from different directions, figure out one's own velocity relative to the aether. For if one were moving through the aether toward the light source, the speed of light should be measured as being faster than if one were at rest (just as water waves would pass you more rapidly if you were swimming toward the source of the waves than if you were floating motionless in the water); whereas if one were moving through the aether away from the light source, the speed of light would be measured as being slower than if one were at rest (just as the water waves would pass you less rapidly if you were swimming away from the source of the waves than if you were floating). Thus, it would be possible to determine experimentally within an inertial frame whether one is at rest in the aether or how fast one is moving through it.

Imagine, then, the consternation when experiments, such as the Michelson-Morley experiment in 1887, failed to detect any motion of the earth through the aether! Despite the fact that the earth is orbiting the sun, the measured speed of light was identical no matter what direction their measuring device was pointed. Some scientists hypothesized that perhaps the earth dragged the aether along with it, rather like an atmosphere, so that the aether seemed to be at rest around the moving earth. But this explanation was ruled out by a well-established phenomenon called the aberration of starlight, which was incompatible with aether drag.

It needs to be underlined how weird the situation was. Waves travel at a constant speed regardless of the motion of their source and in this sense are unlike projectiles, which travel at a velocity which is a combination of the speed of their source plus their speed relative to the source. For example, a bullet fired ahead from a speeding police car travels at a combined speed of the car's speed plus the bullet's normal speed, in contrast to sound waves emitted from the car's siren, which travel through the air at the same velocity whether the car is stationary or in motion. Consequently, an observer who is moving in the same direction as a sound wave will observe it passing him at a slower speed than if he were at rest. If he goes fast enough, he can catch the wave and break the sound barrier. But light waves are different. Light's measured velocity is the *same* in all inertial frames, for *all* observers. This implies, for example, that if an observer in a rocket going 90 percent the speed of light sent a light beam ahead of him, both he and the recipient of the beam would measure the speed of the beam to be the same, and this whether the recipient were standing still or himself moving toward or away from the light source at 90 percent the speed of light.

Desperate for a solution, the Irish physicist George FitzGerald and the great Dutch physicist Hendrick A. Lorentz proposed the remarkable hypothesis that one's measuring devices shrink or contract in the direction of motion through the aether, so that light *appears* to traverse identical distances in identical times, when in fact the distances vary with one's speed. The faster one moves, the more his devices contract, so that the measured speed of light remains constant. Hence, in all inertial frames the speed of light appears the same. With the help of the British scientist Joseph Larmor, Lorentz also came to hypothesize that one's clocks slow down when in motion relative to the aether frame. One thus winds up with Lorentzian relativity: There exists absolute motion, absolute length, and absolute time, but there is no way to discern these experimentally, since motion through the aether affects one's measuring instruments. Lorentz developed a series of equations called the Lorentz transformations, which show how to transform one's own measurements of the spatial and temporal coordinates at which an event occurs into the measurements which would be made by someone in another inertial frame. These transformation equations remain today the mathematical core of STR, even though Lorentz's physical interpretation of STR was different from the most commonly accepted interpretation today.

Hendrick A. Lorentz
"A 'World Spirit' who, not being bound to a specific place, permeated the entire system under consideration and 'in whom' this system existed and who could 'feel' immediately all events would naturally distinguish at once one of the systems U, U', etc., above the others."

In 1905 Albert Einstein, then an obscure clerk in a patent office in Berne, Switzerland, published his own version of relativity. At this time in his young career, Einstein was still a disciple of the great German physicist Ernst Mach. Mach was an ardent empiricist, who detested anything that smacked of metaphysics and who thus sought to reduce statements about entities such as time and space to statements about sense perceptions and the connections between them. The young Einstein took what he called his "epistemological credo" from Mach, holding that knowledge is made up of the totality of sense experiences and the totality of concepts and propositions, which are related in the following way: "The concepts and propositions get 'meaning,' *viz.*, 'content,' only through their connection with sense experience."[8] Any proposition not so connected was, according to Einstein, literally without content, meaningless. Given such a verificationist criterion of meaning, Lorentz's absolute time, space, and motion were "metaphysical" notions and therefore meaningless.

Einstein's 1905 article in the *Annalen der Physik* has been called "the most profoundly revolutionary single paper in the history of physics."[9] He opens his article by jettisoning the aether as superfluous, since, he says, it will not be necessary for the purposes of his paper. In order to talk about motion in a phys-

---

[8] Albert Einstein, "Autobiographical Notes," in *Albert Einstein: Philosopher-Scientist,* Library of Living Philosophers 7 (LaSalle, Ill.: Open Court, 1949), 13.
[9] James T. Cushing, *Philosophical Concepts in Physics* (Cambridge: Cambridge University Press, 1998), 232.

ically meaningful way, Einstein claims, we must be clear about what we mean by "time." Since all judgments about time concern simultaneous events, what we need is a way to determine empirically the simultaneity of distant events. Einstein then proceeds to offer a method of determining, or rather defining, simultaneity for two spatially separated but relatively stationary clocks, that is, two distant clocks sharing the same inertial frame. This procedure will in turn serve as the basis for a definition of the time of an event. He asks us to assume that the time required for light to travel from point A to point B is the same as the time required for light to travel from B to A. Theoretically, light could travel more slowly from A to B and more quickly from B to A, even though the round-trip velocity were always constant. But Einstein says we must assume that the one-way velocity of light is constant. Having made this assumption, he proposes to synchronize clocks at A and B by means of light signals from one to the other. Suppose A sends a signal to B which is in turn reflected back from B to A. If A knows what time it was when he sent the signal to B and what time it was when he received the signal back from B, then he knows that the reading of B's clock when the signal from A arrived was exactly half-way between the time A sent the signal and the time A got the return signal. In this way A and B can arrange to synchronize their clocks. Events are declared to be simultaneous if they occur at the same clock times on synchronized clocks. Using clocks thus synchronized, Einstein defines the time of an event as "the reading simultaneous with the event of a clock at rest and located at the position of the event, this clock being synchronous ... with a specified clock at rest."[10] Now so far, the use of light as the signal plays no special role; one could have used bullets to synchronize distant clocks, so long as the bullets traveled with a uniform velocity.

All this may seem quite unobjectionable and even humdrum. But if you think so, then you have been taken in. The very foundations of the world have just moved! It is with good reason that Banesh Hoffmann advises,

> Watch closely. It will be worth the effort. But be forewarned. As we follow the gist of Einstein's argument, we shall find ourselves nodding in agreement, and later almost nodding in sleep, so obvious and unimportant will it seem. There will come a stage at which we shall barely be able to stifle a yawn. Beware. We shall by then have committed ourselves and it will be too late to avoid the jolt; for the beauty of Einstein's argument lies in its seeming innocence.[11]

---

[10] Albert Einstein, "On the Electrodynamics of Moving Bodies," trans. Arthur Miller in Arthur I. Miller, *Albert Einstein's Special Theory of Relativity* (Reading, Mass.: Addison-Wesley, 1981), 394.
[11] Banesh Hoffmann, cited in Miller, *Einstein's Special Theory of Relativity,* 192.

Albert Einstein
"I allow myself to be deceived as a physicist (and of course the same applies if I am not a physicist) when I imagine that I am able to attach a meaning to the statement of simultaneity."

Its *seeming* innocence! For under the euphemism of disregarding the aether as unnecessary, Einstein thereby abandoned not merely the aether, but, more fundamentally, the aether reference frame, or absolute space. Without absolute space there can be no absolute motion or absolute rest. Bodies are moving or at rest only relative to each other, and it would be meaningless to ask whether an isolated body was stationary or uniformly moving *per se*.

So now suppose that we have inertial frames which are moving with respect to one another, for example, a rocket ship passing near the earth on its way to a distant planet. Suppose that when the rocket ship is close to the earth, its clock agrees with the clock of an earth observer. At that moment the observer on earth sends a light signal to the planet, and an observer on board the rocket does the same. Here the fact that light is the signal plays a crucial role. For since light travels at the same speed relative to all inertial frames, the ship's signal does not travel any faster than the earth's signal, but the two signals travel in tandem and reflect back from the planet together. But in the meantime the rocket ship has moved closer to the planet and so receives the return signal first. Because light's speed is the same for all inertial frames, the observer in the rocket ship cannot detect his own velocity by receiving the signal. The same is true for the earth observer when his signal is then received. But when the rocket and the earth observers divide the light signals' travel times in half, they will get different times for when the signals reached the planet. It might be protested that the rocket ship's measurements are distorted because it was moving toward the planet. But relativity demands that the rocket ship could with equal justice be regarded as at rest, with the planet approaching it and the earth receding away! Remember, on Einstein's theory there is no absolute space and so no absolute rest. Hence, given Einstein's definition of simultaneity, different events are calculated to be simultaneous in different inertial frames, and none of these is the preferred frame giving *the* correct time. All of the various measurements in various frames are correct for each respective frame.

We now see why Einstein entitled his paper, "On the Electrodynamics of Moving Bodies." Given the constancy of the speed of light in all inertial frames, bodies in motion will be related to each other electrodynamically in such a way that the use of electromagnetic signals to establish synchrony relations between them will play havoc with what we normally mean by "simultaneity." What happens is that simultaneity becomes relative. Einstein writes, "Thus we see that we can attribute no *absolute* meaning to the concept of simultaneity, but that two events which, examined from a co-ordinate system, are simultaneous, can no longer be interpreted as simultaneous events when examined from a system which is in motion relatively to that system."[12] What this means is that events which are simultaneous as calculated from one inertial frame will not be simultaneous as calculated from another. An event which lies in *A*'s future may be already present or past for *B*! In fact, events which are not causally connected can even be measured to occur in a different temporal order in different inertial frames!

Einsteinian time and space have many other weird properties, such as time dilation, according to which moving clocks (and all physical processes) run slower and slower as their velocity increases; and length contraction, according to which moving bodies contract in the direction of motion. These were also characteristic of Lorentz's theory, it will be recalled; but the key difference with Einstein's theory is that, since he denies an aether frame, these phenomena are *reciprocal*: For two relatively moving identical rockets *A* and *B*, *B* is shorter than *A* and his clock runs slower than *A*'s relative to *A*'s inertial frame; but *A* is shorter than *B* and his clock runs slower than *B*'s relative to *B*'s inertial frame. Since no inertial frame is preferred, there is no true length or true time *per se*, only lengths and times relative to different frames.

Now, as I said, the Einsteinian world is extraordinarily difficult to conceive. We intuitively think that there is a unique and universal time in which

---

[12] Einstein, "On the Electrodynamics of Moving Bodies," 396.

all events, however distant from one another, occur, and a unique and universal space in which all physical objects exist. But Einstein's theory tells us to substitute for absolute space an infinite number of different spaces, each associated with a different inertial frame, and for absolute time an infinite number of different times, each associated with a different inertial frame. Reality thus is radically fragmented on Einstein's view. Only observers sharing the same inertial frame (that is, at relative rest) have the same time and space. Observers in other inertial frames (that is, in relative motion) live in a different time and space. It is, I think, no exaggeration to say that on Einstein's theory relatively moving observers literally inhabit different worlds which may intersect only at a point. It is no wonder that Einstein's paper is considered revolutionary!

What impact does STR have on the nature of divine eternity? Well, if God is in time, then the obvious question raised by STR is: *Whose time is He in?* For according to Einstein, there is no unique, universal time and so no unique, worldwide "now." Since none of the infinitely many inertial frames is privileged or preferred, no hypothetical observer can justifiably claim that his "now" is the real or true "now." Each inertial frame has its own time and its own present moment, and there is no overarching absolute time in which all these diverse times are integrated into one. So the question is, which is God's "now"?

The defender of divine timelessness maintains that there is no acceptable answer to this question. We cannot plausibly pick out some inertial frame and identify its time as God's time because God is not a physical object in uniform motion, and so the choice of any such frame would be wholly arbitrary. Moreover, it is difficult to see how God, confined to the time of one inertial frame, could be causally sustaining events that are real relative to other inertial frames but are future or past relative to God's frame. Similarly, God's knowledge of what is happening now would be restricted to the temporal perspective of a single frame, leaving Him ignorant of what is actually going on in other frames. In any case, if God were to be associated with a particular inertial frame, then surely, as God's time, the time of that frame would be privileged. It would be the equivalent of the classical aether frame. But then we are back to Lorentzian relativity, not Einsteinian relativity. So long as we maintain, with Einstein, that no frame is privileged, then we cannot identify the time of any inertial frame as God's time.

Nor can we say that God exists in the "now" associated with the time of every inertial frame, for this would obliterate the unity of God's consciousness. In the words of one philosopher of science, "God would have an infinitely split personality, each sub-personality evolving in monad-like isolation from the others"—a hypothesis in which he detects the "faint scent of polytheism."[13] In order to preserve God's consciousness as the consciousness of one being, we must not allow it to be broken and scattered among the inertial frames in the universe.

But if God's time cannot be identified with the time of a single frame or of a plurality of frames, then God must not be in time at all; that is to say, He exists timelessly. We can summarize this reasoning as follows:

1. STR is correct in its description of time.

2. If STR is correct in its description of time, then if God is temporal, He exists in either the time associated with a single inertial frame or the times associated with a plurality of inertial frames.

3. Therefore, if God is temporal, He exists in either the time associated with a single inertial frame or the times associated with a plurality of inertial frames.

4. God does not exist in either the time associated with a single inertial frame or the times associated with a plurality of inertial frames.

5. Therefore, God is not temporal.

Critique

What can be said in response to this argument? Although it may come as something of a shock to many, it seems to me that the most dubious premise of the above reasoning is premise (1). In order to understand why I say this, let us recur to Newton's distinction between absolute and relative time. While it is easy to find statements by prominent physicists and philosophers to the effect that STR destroyed the concept of absolute time and so forces us to abandon the classical concept of time, such verdicts are almost invariably based upon a superficial understanding of the metaphysical foundations of Newton's doctrine of absolute time (and space).

We have already seen that Newtonian time is absolute both in the sense that time itself is distinct from our measures of time and in the sense that there

---

[13] Paul Fitzgerald, "Relativity Physics and the God of Process Philosophy," *Process Studies* 2 (1972): 259, 260.

is a unique, all-embracing time. But as is well-known, Newton also conceived of time as absolute in yet a third, more profound sense, namely, he held that time exists independently of any physical objects whatsoever. Usually, this is interpreted to mean that time would exist even if nothing else existed, that we can conceive of a logically possible world which is completely empty except for the container of absolute space and the flow of absolute time.

But here we must be very careful. Modern secular scholars tend frequently to forget how ardent a theist Newton was and how central a role this theism played in his metaphysical outlook. Noting that Newton considered God to be temporal and therefore time to be everlasting, David Griffin observes that, "Most commentators have ignored Newton's heterodox theology, and his talk of 'absolute time' has been generally misunderstood to mean that time is not in any sense a relation and hence can exist apart from actual events."[14] In fact, Newton makes quite clear in the *General Scholium* to the *Principia,* which he added in 1713, that absolute time and space are constituted by the divine attributes of eternity and omnipresence. He writes,

> He is eternal and infinite . . . ; that is, his duration reaches from eternity to eternity; his presence from infinity to infinity. . . . He is not eternity and infinity, but eternal and infinite; he is not duration or space, but he endures and is present. He endures forever, and is everywhere present; and, by existing always and everywhere, he constitutes duration and space. Since every particle of space is *always,* and every indivisible moment of duration is *everywhere,* certainly the Maker and Lord of all things cannot be *never* and *nowhere.*[15]

Because God is eternal, there exists an everlasting duration, and because He is omnipresent, there exists an infinite space. Absolute time and space are therefore relational in that they are contingent upon the existence of God.

In his earlier treatise, "On the Gravity and Equilibrium of Fluids," Newton argued that space (and by implication time) is neither a substance, nor a property, nor nothing at all. It cannot be nothing because it has properties, such as infinity and uniformity in all directions. It cannot be a property because it can exist without bodies. Neither is it a substance: "It is not substance . . . because it is not absolute in itself, but is as it were an emanant

---

[14] David Ray Griffin, "Introduction: Time and the Fallacy of Misplaced Concreteness," in *Physics and the Ultimate Significance of Time,* ed. David R. Griffin (Albany, N.Y.: State University of New York Press, 1986), 6-7.
[15] Newton, *Principles of Natural Philosophy,* 2:545.

effect of God, or a disposition of all being. . . ."[16] Contrary to the conventional understanding, Newton here declares explicitly that space is *not* in itself absolute and therefore not a substance. Rather it is an emanent—or emanative—effect of God. By this notion Newton meant to say that time and space were the immediate consequence of God's very being. God's infinite being has as its consequence infinite time and space, which represent the quantity of His duration and presence. Newton does not conceive of space or time as in any way attributes of God Himself, but rather, as he says, concomitant effects of God.

In Newton's view God's "now" is thus the present moment of absolute time. Since God is not "a dwarf-god" located at a particular place in space,[17] but is omnipresent, there is a worldwide moment which is absolutely present. Newton's temporal theism thus provides the foundation for absolute simultaneity. The absolute present and absolute simultaneity are features first and foremost of God's time, absolute time, and derivatively of measured or relative time.

Thus, the classical, Newtonian concept of time is firmly rooted in a theistic worldview. What Newton did not realize, nor could have suspected, is that physical time is not only *relative* but also *relativistic,* that the approximation of physical time to absolute time depends not merely upon the regularity of one's clock but also upon its motion. Unless a clock were at absolute rest, it would not accurately register the passage of absolute time. Moving clocks run slowly. This truth, unknown to Newton, was finally grasped by scientists only with the advent of Relativity Theory.

Where Newton fell short, then, was not in his analysis of absolute or metaphysical time—he had theological grounds for positing such a time—but in his incomplete understanding of relative or physical time. He assumed too readily that an ideal clock would give an accurate measure of time independently of its motion. If confronted with relativistic evidence, Newton would no doubt have welcomed this correction and seen therein no threat at all to his doctrine of absolute time.[18] In short, relativity corrects Newton's concept of physical time, not his concept of absolute time.

Of course, it hardly needs to be said that there is a great deal of antipathy

---

[16] Isaac Newton, "On the Gravity and Equilibrium of Fluids," [*De gravitatione et aequipondio fluidorum*], in *Unpublished Scientific Papers of Isaac Newton,* ed. A. Rupert Hall and Marie Boas Hall (Cambridge: Cambridge University Press, 1962), 132.

[17] Isaac Newton, "Place, Time, and God," in J. E. McGuire, "Newton on Place, Time, and God: An Unpublished Source," *British Journal for the History of Science* 11 (1978): 123.

[18] John Lucas emphasizes, "The relativity that Newton rejected is not the relativity that Einstein propounded; and although the Special Theory of Relativity has shown Newton to be wrong in some respects, . . . it has not shown that time is relative in Newton's sense, and merely some numerical measure of process" (J. R. Lucas, *A Treatise on Time and Space* [London: Methuen, 1973], 90).

in modern physics and philosophy of science toward such metaphysical realities as Newtonian space and time, primarily because they are not physically detectable. But Newton would have been singularly unimpressed with this verificationist equation between physical undetectability and non-existence. The grounds for metaphysical space and time were not physical but philosophical, or more precisely, theological. Epistemological objections fail to worry Newton because, as Oxford philosopher John Lucas nicely puts it, "He is thinking of an omniscient, omnipresent Deity whose characteristic relation with things and with space is expressed in the imperative mood."[19] Modern physical theories say nothing against the existence of such a God or against the metaphysical time constituted, in Newton's thinking, by His eternity. What relativity theory did, in effect, was simply to remove God from the picture and to substitute in His place a finite observer. "Thus," according to historian of science Gerald Holton, "the RT [Relativity Theory] merely shifted the focus of space-time from the sensorium of Newton's God to the sensorium of Einstein's abstract *Gedankenexperimenter*—as it were, the final secularization of physics."[20] But to a man like Newton, such a secular outlook impedes rather than advances our understanding of the nature of reality.

What Einstein did, in effect, was to shave away Newton's absolute time and space, and along with them the aether, thus leaving behind only their empirical measures. Since these are relativized to inertial frames, one ends up with the relativity of simultaneity and of length. What justification did Einstein have for so radical a move? How did he know that absolute time and space do not exist? The answer, in a word, is verificationism. According to verificationism, statements which cannot be in principle empirically verified are meaningless. Historians of science have demonstrated convincingly that at the philosophical roots of Einstein's theory lies a verificationist epistemology, mediated to the young Einstein chiefly through the influence of Ernst Mach, which comes to expression in Einstein's analysis of the concepts of time and space.[21]

---

[19] Ibid., 143.
[20] Gerald Holton, "On the Origins of the Special Theory of Relativity," in Gerald Holton, *Thematic Origins of Scientific Thought: Kepler to Einstein* (Cambridge, Mass.: Harvard University Press, 1973), 171. The sensorium was conceived to be that aspect of the mind in which mental images of physical objects are formed. Newton said that because physical objects exist in space and God is omnipresent, they literally exist in God and thus are immediately present to Him. Absolute space is, as it were, God's sensorium in the sense that He has no need of mental images of things, since the things themselves are present to Him. Einstein's *Gedankenexperimenter* (thought experimenter) is the hypothetical observer associated with any inertial frame, for whom time and space are purely relative quantities.
[21] See especially Gerald J. Holton, "Mach, Einstein and the Search for Reality," in *Ernst Mach: Physicist and Philosopher*, Boston Studies in the Philosophy of Science 6 (Dordrecht: D. Reidel, 1970), 165-199; idem, "Where Is Reality? The Answers of Einstein," in *Science and Synthesis*, ed. UNESCO (Berlin: Springer-Verlag, 1971), 45-69; and the essays collected together in idem, *Thematic Origins of Scientific Thought*.

The introductory sections of Einstein's 1905 paper are predicated squarely upon verificationist assumptions. These come through most clearly in his operationalist redefinition of key concepts. Einstein proposes to define concepts such as *time* and *simultaneity* in terms of empirically verifiable operations. The meaning of "time" is made to depend upon the meaning of "simultaneity," which is defined locally in terms of occurrence at the same local clock reading. In order to define a common time for spatially separated clocks, we adopt the convention that the time which light takes to travel from A to B equals the time it takes to travel from B to A—a definition that *presupposes* that absolute space does not exist. For if A and B are at relative rest but moving in tandem through absolute space, then it is not the case that a light beam will travel from A to B in the same amount of time it takes to travel from B to A, since the distances traversed will not be the same (Fig. 2.1).

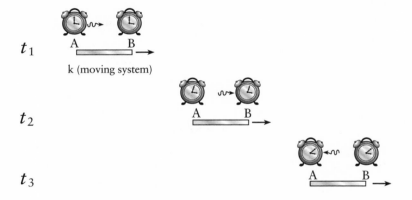

Fig. 2.1: Clock synchronization of relatively stationary clocks in absolute motion. A light signal is first sent from A toward B. By the time the signal reaches B, both A and B will have moved together some distance from the point where A first released the signal. Finally, by the time the reflected signal from B reaches A again, both A and B will have moved still farther from the release point. Since the signal traveled farther from A to B than from B back to A, the time it took to travel from A to B is greater than the time it took to travel from B to A.

That is why Einstein's theory, far from *disproving* the existence of absolute space, actually *presupposes* its non-existence. All of this is done by mere stipulation. Reality is reduced to what our measurements read; Newton's metaphysical time and space, which transcend operational definitions, are assumed to be mere figments of our imagination.

How, then, shall we assess the claim that STR has eliminated absolute time and space? The first thing to be said is that the verificationism which characterized Einstein's original formulation of STR belongs essentially to the philosophical foundations of the theory. The whole theory rests upon Einstein's redefinition of simultaneity in terms of clock synchronization by light signals. But that redefinition assumes necessarily that the time that light takes to travel between two relatively stationary observers A and B is the same from A to B as from B to A in a round-trip journey. That assumption presupposes that A and B are not both in absolute motion, or in other words that neither absolute space nor a privileged inertial frame exists. The only justification for that assumption is that it is empirically impossible to distinguish uniform motion from rest relative to such a frame, and if absolute space and absolute motion or rest are undetectable empirically, they do not exist (and may even be said to be meaningless).

In a clear-sighted analysis of the epistemological foundations of STR, University of Michigan philosopher of science Lawrence Sklar underlines the essential role played by this verificationism:

> Certainly the original arguments in favor of the relativistic viewpoint were rife with verificationist presuppositions about meaning, etc. And despite Einstein's later disavowal of the verificationist point of view, no one to my knowledge has provided an adequate account of the foundations of relativity which isn't verificationist in essence.[22]

It would be desirable to do so, muses Sklar, but "what I don't know is . . . how to formulate a coherent underpinning for relativity which isn't verificationist. . . ."[23]

But if verificationism belongs essentially to the foundations of STR, the next thing to be said is that verificationism has proved to be completely untenable and is now outmoded. The untenability of verificationism is so universally acknowledged that it will not be necessary to rehearse the objections against it here.[24] Richard Healey observes that verificationism "has come

---

[22] Lawrence Sklar, "Time, Reality, and Relativity," in *Reduction, Time and Reality,* ed. Richard Healey (Cambridge: Cambridge University Press, 1981), 141.
[23] Ibid.
[24] See the excellent survey in Frederick Suppe, "The Search for Philosophical Understanding of Scientific Theories," in *The Structure of Scientific Theories,* 2d ed., ed. F. Suppe (Urbana, Ill.: University of Illinois Press, 1977), 3-118. Verificationism was far too restrictive a theory of meaning to be plausible, for it would force us to dismiss as meaningless vast tracts of human discourse, including not just metaphysical and theological statements but also aesthetic and ethical statements, as well as many scientific statements (e.g., the postulate of the constancy of the one-way velocity of light, an unprovable assumption which lies at the heart of STR). Worse, verificationism turned out to be self-refuting. For the statement "Only sentences which can in principle be empirically verified are meaningful" is itself not an empirically verifiable sentence and so is by its own standard meaningless!

under such sustained attack that opposition to it has become almost orthodoxy in the philosophy of science."[25] Verificationism provides no justification for thinking that Newton erred, for example, in holding that God exists in a time which exists independently of our physical measures of it and which may or may not be accurately registered by them. It matters not a whit whether we finite creatures know what time it is in God's absolute time; God knows, and that is enough.

Contemporary physics has in any case ignored the constrictions of verificationism. When the contemporary student of physics reads the anti-metaphysical polemics of the past generation, he must feel as though he were peering into a different world! George Gale, in surveying some of "the metaphysical perplexities abounding in today's physics," contends, "... we are entering a phase of scientific activity during which the physicist has out-run his philosophical base camp, and, finding himself cut off from conceptual supplies, he is ready and waiting for some relief from his philosophical comrades-in-arms."[26] In various fields such as quantum mechanics, classical cosmology, and quantum cosmology, debates rage over issues that are overtly metaphysical in character. Take quantum mechanics, for example. One recent expert has said, "In an effort to understand the quantum world, we are led beyond physics, certainly into philosophy and maybe even into cosmology, psychology and theology."[27] Cosmology has become grandly metaphysical. "Cosmology, even as practiced today," says Gale, "is science done at the limit: at the limit of our concepts, of our mathematical methods, of our instruments, indeed, of our very imaginations.... it is clear that metaphysics continues to play an honorable role in cosmology. And, to the extent that it is an honorable role, it is no dishonor to use metaphysics in one's cosmologizing."[28] Noting that in recent years such "metaphysical conundrums" as *creatio ex nihilo* "have entered the mainstream of scientific discussions," John Barrow remarks, "Traditional dogmas as to what criteria must be met by a body of ideas for it to qualify as a 'science' now seem curiously inappropriate in the face of problems and studies far removed from the human enterprise."[29] The verificationist, anti-metaphysical view of physics which dominated the first two-thirds of the twentieth century is simply outmoded in light of contemporary theoretical physics.

---

[25] Richard Healey, "Introduction," in Healey, ed., *Reduction, Time and Reality*, vii.
[26] George Gale, "Some Metaphysical Perplexities in Contemporary Physics," paper presented at the 36th Annual Meeting of the Metaphysical Society of America, Vanderbilt University, March 14-16, 1985.
[27] Euan Squires, *The Mystery of the Quantum World* (Bristol: Adam Hilger, 1986), 4.
[28] George Gale, "Cosmos and Conflict," paper presented at the conference "The Origin of the Universe," Colorado State University, Ft. Collins, Colorado, September 22-25, 1988.
[29] John Barrow, *The World within the World* (Oxford: Clarendon, 1988), 2, vii-viii.

It is difficult, therefore, to understand how many contemporary philosophers and physicists can still speak of STR's "forcing" us to abandon the classical concepts of space and time or of STR's "destruction" of Newtonian absolute time. Lawrence Sklar concludes,

> The original Einstein papers on special relativity are founded, as is well known, on a verificationist critique of earlier theories. . . . Now it might be argued that Einstein's verificationism was a misfortune, to be encountered not with a rejection of special relativity, but with an acceptance of the theory now to be understood on better epistemological grounds. . . .
>
> But I don't think a position of this kind will work in the present case. I can see no way of rejecting the old aether-compensatory theories, originally invoked to explain the Michelson-Morley results, without invoking a verificationist critique of some kind or other.[30]

With the demise of verificationism, the philosophical underpinnings of STR have collapsed. In short, there is no reason to think that premise

1. STR provides the correct description of time

is true.

Now let me be very clear that I am not here endorsing Newton's view on divine eternity; but I am saying that the theologian who, like Newton, believes God to be temporal need not feel threatened by STR, because STR's claim that absolute time does not exist is founded essentially upon a defunct and untenable epistemology.

If we do suppose that God is in time, how then should we understand STR? Henri Poincaré, the great French mathematician and precursor of STR, helped to point the way. In a fascinating passage in his essay "The Measure of Time," Poincaré briefly entertains the hypothesis of "an infinite intelligence" and considers the implications of such a hypothesis. Poincaré is reflecting on the problem of how we can apply one and the same measure of time to spatially distant events. What does it mean, for example, to say that two thoughts in two people's minds occur simultaneously? Or what does it mean to say that a supernova occurred before Columbus saw the New World? Like a good verificationist, Poincaré says, "All these affirmations have by themselves no meaning."[31] Then he remarks,

---

[30] Sklar, "Time, Reality and Relativity," 132.
[31] Henri Poincaré, "The Measure of Time," in *The Foundations of Science,* trans. G. B. Halstead (Science Press: 1913; rep. ed.: Washington, D.C.: University Press of America, 1982), 228.

> We should first ask ourselves how one could have had the idea of putting into the same frame so many worlds impenetrable to one another. We should like to represent to ourselves the external universe, and only by so doing could we feel that we understood it. We know we can never attain this representation: our weakness is too great. But at least we desire the ability to conceive an infinite intelligence for which this representation could be possible, a sort of great consciousness which should see all, and which should classify all *in its time*, as we classify, *in our time*, the little we see.
>
> This hypothesis is indeed crude and incomplete, because this supreme intelligence would be only a demigod; infinite in one sense, it would be limited in another, since it would have only an imperfect recollection of the past; it could have no other, since otherwise all recollections would be equally present to it and for it there would be no time. And yet when we speak of time, for all which happens outside of us, do we not unconsciously adopt this hypothesis; do we not put ourselves in the place of this imperfect God; and do not even the atheists put themselves in the place where God would be if he existed?
>
> What I have just said shows us, perhaps, why we have tried to put all physical phenomena into the same frame. But that cannot pass for a definition of simultaneity, since this hypothetical intelligence, even if it existed, would be for us impenetrable. It is therefore necessary to seek something else.[32]

Poincaré here suggests that, in considering the notion of simultaneity, we instinctively put ourselves in the place of God and classify events as past, present, or future according to His time. Poincaré does not deny that from God's perspective there would exist relations of absolute simultaneity. But he rejects the hypothesis as yielding a definition of simultaneity because we could not know such relations; such knowledge would remain the exclusive possession of God Himself.

Clearly, Poincaré's misgivings are relevant to a definition of simultaneity only if one is presupposing some sort of verificationist theory of meaning, as he undoubtedly was. The fact remains that God knows the absolute simultaneity of events even if we grope in total darkness. Nor need we be concerned with Poincaré's argument that such an infinite intelligence would be a mere demigod, since there is no reason to think that a temporal being cannot have a perfect recollection of the past. There is no conceptual difficulty in the idea of a being that knows all past-tense truths. His knowledge would be constantly changing, as more and more events become past. But at each succes-

---
[32] Ibid., 228-229.

sive moment he could know every past-tense truth that there is at that moment. Hence, it does not follow that if God is temporal, He cannot have perfect recollection of the past.

Poincaré's hypothesis suggests, therefore, that if God is temporal, His present is constitutive of relations of absolute simultaneity.[33] On this view, the philosopher J. M. Findlay was wrong when he said, "the influence which harmonizes and connects all the world-lines is not God, not any featureless, inert, medium, but that living, active interchange called . . . Light, offspring of Heaven firstborn."[34] On the contrary, the use of light signals to establish clock synchrony would be a convention which finite and ignorant creatures have been obliged to adopt, but the living and active God, who knows all, would not be so dependent. Inviting us to "imagine a superhuman observer— a god—who is not bound by the limitations of the maximum velocity of light," Milton K. Munitz notes,

> Such an observer could survey in a single instant the entire domain of galaxies that have already come into existence. His survey would not have to depend on the finite velocity of light. It would not betray any restriction in information of the kind that results from the delayed time it takes to bring information about the domain of galaxies to an ordinary human observer situated in the universe, and who is therefore bound by the mechanisms and processes of signal transmission. The entire domain of galaxies would be seen instantaneously by this privileged superhuman observer. His observational survey of all galaxies would yield what Milne calls a "world map."[35]

In God's temporal experience, there would be a moment which would be present in absolute time, whether or not it were registered by any clock time. He would know, without any dependence on clock synchronization proce-

---

[33] Cf. Lorentz's illustration in a letter to Einstein in January of 1915 in response to the latter's paper "The Formal Foundations of the General Theory of Relativity." In a passage redolent of the *General Scholium* and *Opticks* of Newton, Lorentz broached considerations whereby "I cross the borderland of physics":
"A 'World Spirit' who, not being bound to a specific place, permeated the entire system under consideration or 'in whom' this system existed and who could 'feel' immediately all events would naturally distinguish at once one of the systems U, U', etc. above the others" (H. A. Lorentz to A. Einstein, January, 1915, Boerhaave Museum, cited in Jozsef Illy, "Einstein Teaches Lorentz, Lorentz Teaches Einstein. Their Collaboration in General Relativity, 1913–1920," *Archive for History of Exact Sciences* 39 [1989]: 274).
Such a being, says Lorentz, could "directly verify simultaneity."

[34] J. M. Findlay, "Time and Eternity," *Review of Metaphysics* 32 (1978–1979): 6-7.

[35] Milton K. Munitz, *Cosmic Understanding* (Princeton: Princeton University Press, 1986), 157. Kanitscheider concludes that only an omnipresent, cosmic observer who sees the world *sub specie aeternitatis* (from the perspective of eternity) can be in the position to draw up a world map (Bernulf Kanitscheider, *Kosmologie* [Stuttgart: Philipp Reclam, Jun., 1984], 193).

dures or any physical operations at all, which events were simultaneously present in absolute time. He would know this simply in virtue of His knowing at every such moment the unique set of present-tense truths at that moment, without any need of physical observation of the universe.

So what would become of STR if God is in time? From what has been said, God's existence in time would imply that Lorentz, rather than Einstein, had the correct interpretation of Relativity Theory. That is to say, Einstein's clock synchronization procedure would be valid only in the preferred (absolute) reference frame, and measuring rods would contract and clocks slow down in the customary special relativistic way when in motion with respect to the preferred frame. Such an interpretation would be implied by divine temporality, for God in the "now" of absolute time would know which events in the universe are now being created by Him and are therefore absolutely simultaneous with each other and with His "now." This startling conclusion shows that Newton's theistic hypothesis is not some idle speculation but has important implications for our understanding of how the world is and for the assessment of rival scientific theories.

Lorentzian relativity is admitted on all sides to be empirically equivalent to Einsteinian relativity, and there are even indications on the cutting edge of science today that a Lorentzian view may be preferable in light of recent discoveries. In fact, due to developments in quantum physics (the physics of the subatomic realm), there has been what one participant in the debate has called a "sea change" in the attitude of the physics community toward Lorentzian relativity.[36]

For example, the best explanation of the experimental evidence concerning what is called Bell's Theorem seems to be that relations of absolute simultaneity do exist. First a bit of background: Between 1927 and 1935 Einstein pursued a running argument with the Danish physicist Niels Bohr, the father of quantum physics. Bohr believed that elementary particles do not have intrinsic, determinate properties such as momentum and location. Such properties are possessed only in relation to some measuring apparatus. Retorting that "God does not play dice," Einstein repeatedly sought to concoct thought experiments which would show that, contrary to Bohr, the subatomic world is not characterized by indeterminacy. The most celebrated of these was a thought experiment jointly proposed with Boris Podolsky and

---

[36] John Kennedy in a paper delivered to the American Philosophical Association, Central Division Meeting, Pittsburgh, Pa., April 23-26, 1997. Compare the passing remark of Balashov, "the idea of restoring absolute simultaneity no longer has a distinctively pseudo-scientific flavor it has had until very recently" (Yuri Balashov, "Enduring and Perduring Objects in Minkowski Space-Time," *Philosophical Studies* 99 (2000): 159).

Nathan Rosen in 1935, which thus came to be known as the EPR experiment.[37] The idea was to split a beam of light into two particles traveling in opposite directions. If we measure the velocity of one of the particles, quantum physics requires that the other particle instantaneously take on a similar value. Since no causal influence can travel faster than the speed of light, there is no way in which our measuring one particle could influence the other particle. Thus, the particles must possess an intrinsic, determinate velocity even before they are measured—in contradiction to Bohr's claim. In 1964 John Bell showed that if Einstein were right, then such an experiment would have testable consequences which disagree with the predictions made by quantum theory.[38] Tests were run, and, lo and behold, the predictions of quantum physics were fully vindicated.

The implications were enormous. In order to explain the results, one must either postulate faster-than-light causal influences between the particles or hold that the particles are somehow non-causally correlated so that both particles instantaneously take on certain determinate properties. In either case, the relativity of simultaneity posited by STR will have to be given up. Bell himself, pondering the implications of the experimental data, mused,

> I think it's a deep dilemma, and the resolution of it will not be trivial; it will require a substantial change in the way we look at things. But I would say that the cheapest resolution is something like going back to relativity as it was before Einstein, when people like Lorentz and Poincaré thought that there was an aether—a preferred frame of reference—but that our measuring instruments were distorted by motion in such a way that we could not detect motion through the aether. . . . that is certainly the cheapest solution. Behind the apparent Lorentz invariance of the phenomena, there is a deeper level which is not Lorentz invariant. . . . what is not sufficiently emphasized in textbooks, in my opinion, is that the pre-Einstein position of Lorentz and Poincaré, Larmor and Fitzgerald was perfectly coherent, and is not inconsistent with relativity theory. The idea that there is an aether, and these Fitzgerald contractions and Larmor dilations occur, and that as a result the instruments do not detect motion through the aether—that is a perfectly coherent point of view. . . . The reason I want to go back to the idea of an aether here is because in these *EPR* experiments there is the suggestion that

---

[37] A. Einstein, B. Podolsky, and N. Rosen, "Can Quantum Mechanical Description of Physical Reality Be Considered Complete?" reprinted in *Quantum Theory and Measurement*, ed. John Archibald Wheeler and Wojciech Hubert Zurek, Princeton Series in Physics (Princeton: Princeton University Press, 1983), 138.

[38] J. S. Bell, "On the Einstein Podolsky Rosen Paradox," reprinted in *Quantum Theory and Measurement*, 403-408.

behind the scenes something is going faster than light. Now if all Lorentz frames are equivalent, that also means that things can go backward in time. . . . [This] introduces great problems, paradoxes of causality, and so on. And so it is precisely to avoid these that I want to say there is a real causal sequence which is defined in the aether.[39]

In light of the above, it is little wonder that the great philosopher of science Karl Popper considered the experiments run on Bell's Theorem as the first crucial experiments between Lorentz's and Einstein's interpretations of relativity. He remarks,

> The reason for this assertion is that the mere existence of an infinite velocity entails that of an absolute simultaneity and thereby of an absolute space. Whether or not an infinite velocity can be attained *in the transmission of signals* is irrelevant for this argument: the one inertial system for which Einsteinian simultaneity coincides with absolute simultaneity . . . would be the system at absolute rest—whether or not this system at absolute rest can be experimentally identified.[40]

If there is action at a distance, advises Popper, "it would mean that we have to give up Einstein's interpretation of special relativity and return to Lorentz's interpretation and with it to Newton's absolute space and time."[41] Popper goes on to observe that none of the mathematical formalism of STR need be given up, but only Einstein's interpretation of it. "If we now have theoretical reasons from quantum theory for introducing absolute simultaneity, then we would have to go back to Lorentz's interpretation."[42]

Moreover, in a truly astonishing development in twentieth-century cosmology, we may even have a good idea as to what is the preferred reference frame. For the cosmic microwave background radiation first predicted by George Gamow and then discovered in 1965 by A. A. Penzias and R. W. Wilson is at rest with respect to the expanding space of Big Bang cosmology.

---

[39] "John Bell," interview in P. C. W. Davies and J. R. Brown, *The Ghost in the Atom* (Cambridge: Cambridge University Press, 1986), 45-47. Even if one does not postulate faster-than-light causal influences, the fact remains that the indeterminacy in each particle collapses instantly and simultaneously, which cannot be accounted for within an Einsteinian interpretation of STR, as is lucidly explained by Tim Maudlin, *Quantum Non-Locality and Relativity*, Aristotelian Society Series 13 (Oxford: Blackwell, 1994).
[40] Karl Popper, "A Critical Note on the Greatest Days of Quantum Theory," in *Quantum, Space and Time—The Quest Continues*, ed. Asim O. Barut, Alwyn van der Merwe, and Jean-Pierre Vigier, Cambridge Monographs on Physics (Cambridge: Cambridge University Press, 1984), 54.
[41] Karl Popper, *Quantum Theory and the Schism in Physics*, ed. W. W. Bartley III (Totowa, N.J.: Rowman and Littlefield, 1982), 29.
[42] Ibid., 30.

It is therefore a sort of aether, serving to distinguish a universal rest frame.[43] Recent tests have even detected the earth's motion relative to this background radiation, thus fulfilling the dream of nineteenth-century physics of measuring the aether wind![44] What nineteenth-century physics could not detect using visible light radiation, twentieth-century physics has discovered using microwave radiation. Philosopher of science James Cushing connects the universal preferred frame defined by the microwave background radiation with the unique frame in which absolute simultaneity is required by the experimental results of Bell's Theorem, proclaiming, "Today . . . the aether has re-emerged through quantum phenomena!"[45] One can only speculate whether, had these facts been known in 1905, Einstein would have ever suggested that absolute space and time do not exist.

Again, none of this proves that Newton was right in thinking that God is in time; but it does undercut the claim that STR has proven Newton to be wrong. The defender of divine temporality can plausibly reject the first premise of the argument for divine timelessness based on the Special Theory of Relativity.

But what about the second premise?

2. If STR is correct in its description of time, then if God is temporal, He exists in either the time associated with a single inertial frame or the times associated with a plurality of inertial frames.

Is this premise true? The difficulty with this premise is that it fails to take into account the fact that STR is a *restricted* theory of relativity and therefore is correct only within prescribed limits. It is a theory which deals with uniform motion only. The analysis of non-uniform motion, such as acceleration and rotation, is provided by the General Theory of Relativity (GTR). STR cannot therefore be expected to give us the final word about the nature of time and space; indeed, within the context of GTR a new and important conception of time emerges.

---

[43] Michael Heller, Zbigniew Klimek, and Konrad Rudnicki, "Observational Foundations for Assumptions in Cosmology," in *Confrontation of Cosmological Theories with Observational Data*, ed. M. S. Longair (Dordrecht: D. Reidel, 1974), 4. Kanitscheider remarks, "The cosmic background radiation thereby furnishes a reference frame, relative to which it is meaningful to speak of an *absolute motion*" (Kanitscheider, *Kosmologie*, 256).

[44] G. F. Smoot, M. Y. Gorenstein, and R. A. Muller, "Detection of Anisotropy in the Cosmic Blackbody Radiation," *Physical Review Letters* 39 (1977): 899.

[45] James T. Cushing, "What Measurement Problem?" in *Perspectives on Quantum Reality*, ed. Rob Clifton, University of Western Ontario Series in Philosophy of Science 57 (Dordrecht: Kluwer Academic Publishers, 1996), 75. So also Popper, *Quantum Theory*, 30.

Let us therefore offer a brief word of explanation of GTR. As in Newtonian physics, so in STR accelerated or rotational motion is not relative but absolute. If a reference frame is accelerating, rather than moving uniformly, there are discernible effects within the frame. For example, a space traveler feels himself pressed back into his seat as his rocket accelerates; by the same token, deceleration causes him to pitch forward in his seat. Troubled by the non-equivalence of inertial and non-inertial frames, Einstein endeavored in his GTR to enunciate a General Principle of Relativity which would serve to render physically equivalent all inertial and non-inertial frames alike. In his article "The Foundations of General Relativity Theory" (1915), he boasted that his theory "takes away from space and time the last remnant of physical objectivity."[46] It was, in effect, intended to be the final destruction of Newton's absolute space and time.

What Einstein saw was that the effects of acceleration were exactly equivalent to the effects of gravitation. A space traveler in a capsule suddenly pressed back into his seat would not know the difference between his rocket's accelerating or an increase in the force of gravity behind him. (Hence, today we often speak of an astronaut's feeling a force of several G's [several times the normal force of gravity] as his rocket blasts off.) Perhaps, then, gravity and acceleration could be regarded as equivalent.

In order to carry out this idea, Einstein proposed that gravity be analyzed, not as a force that somehow affects objects at a distance, but rather as the acceleration of objects in space-time. A physical object bends or warps space-time, just as a heavy ball resting on a cushion warps the cushion, so that objects that appear to be under another object's gravitational influence are not in fact being *pulled* toward the larger object, but are rather, to put it crudely, coasting downhill toward it. A two-dimensional analogy would be a taut rubber sheet with balls of various mass placed on the sheet, causing depressions around them of various depths. If a ball bearing were rolled across the sheet and hit a depression, it would be deflected from its path and maybe even "pulled" into the object causing the depression. Although the three-dimensional analogue of this is not visualizable, Einstein worked out an incredibly complex mathematical theory for it which came to replace Newton's theory of gravitation.

In fact, however, Einstein was only partially successful in achieving his aims. He did not succeed in enunciating a tenable General Principle of Relativity, nor was he able to show the physical equivalence of all reference

---

[46] A. Einstein, "The Foundations of General Relativity Theory," in *General Theory of Relativity*, ed. C. W. Kilminster, Selected Readings in Physics (Oxford: Pergamon Press, 1973), 148.

frames.[47] Acceleration and rotation are still distinguishable from uniform motion in the context of GTR. He did succeed in drafting a revolutionary and complex theory of gravitation, which has been widely hailed as his greatest intellectual achievement. The so-called General Theory of Relativity is thus something of a misnomer: It is really a theory of gravitation and not an extension of the Special Theory of Relativity from inertial reference frames to all reference frames.

It might appear, therefore, that GTR has nothing more to contribute to our understanding of time than STR. The two theories appear to differ simply over whether space-time is curved; if one adds a condition of flatness of space-time to GTR, then STR results. Such a conclusion would be mistaken, however. For GTR serves to introduce into Relativity Theory a cosmic perspective, enabling us to draft cosmological models of the universe governed by the gravitational field equations of GTR. Within the context of such cosmological models, the issue of time resurfaces dramatically.

Einstein himself proposed the first GTR-based cosmological model in his paper, "Cosmological Considerations on the General Theory of Relativity," in 1917.[48] The model describes a universe whose temporal dimension is infinite but whose spatial dimensions are finite and invariable. Thus, four-dimensional space-time has the form of a cylinder, time represented by the length of the cylinder and space by its cross-sections. The German philosopher of science Bernulf Kanitscheider draws our attention to the time coordinate which shows up in Einstein's model:

> It represents in a certain sense the restoration of the universal time which was destroyed by STR. In the static world there is a global reference frame, relative to which the whole of cosmic matter finds itself at rest. All cosmological parameters are independent of time. In the rest frame of cosmic matter space and time are separated. For fundamental observers at rest, all clocks can be synchronized and a worldwide simultaneity can be defined in this cosmic frame.[49]

---

[47] See Michael Friedman, *Foundations of Space-Time Theories* (Princeton: Princeton University Press, 1983), 204-215; also Hermann Bondi, "Is 'General Relativity' Necessary for Einstein's Theory of Gravitation?" in *Relativity, Quanta, and Cosmology in the Development of the Scientific Thought of Albert Einstein,* ed. Francesco De Finis, 2 vols. (New York: Johnson Reprint Corp., 1979), 179-186.
[48] Albert Einstein, "Cosmological Considerations on the General Theory of Relativity," in *The Principle of Relativity,* by Albert Einstein, et al., with notes by A. Sommerfeld, trans. W. Perrett and J. B. Jeffery (reprint, New York: Dover Publications, 1952), 177-188.
[49] Kanitscheider, *Kosmologie,* 155. See also G. J. Whitrow, *The Natural Philosophy of Time,* 2d ed. (Oxford: Clarendon, 1980), 283-284.

Thus, cosmological considerations prompt the conception of a cosmic time which measures the duration of the universe as a whole.

Nor is this cosmic time limited to Einstein's model of a static universe. Models of an expanding universe, which trace their origin to Willem de Sitter's 1917 model,[50] may also involve a cosmic time. All contemporary expansion models derive from Russian physicist Alexander Friedman's 1922 model of an expanding, material universe characterized by ideal homogeneity and uniformity.[51] Several features of the cosmic time in Friedman models merit comment. First, although one may slice space-time into various spatial cross-sections wholly arbitrarily, certain space-times have natural symmetries that guide the construction of cosmic time.[52] GTR does not itself mandate any formula for how to slice up space-time; it has no inherent "layers." Theoretically, then, one may slice it up at one's whim. Nevertheless, certain models of space-time, like the Friedman model, have a dynamical, evolving spatial geometry, and in order to ensure a smooth development of this geometry, it will be necessary to construct a time parameter based on a preferred slicing of space-time.

To borrow an illustration from Sir Arthur Eddington, we may think of space-time on the analogy of either a stack of paper or a solid block of paper. The solid block could theoretically be sliced in any way into a series of sheets. But suppose that on each page in the stack of paper, there is drawn a cartoon figure, such that by flipping through the pages successively, one sees the figure animated into action. Any other slicing of the block would result merely in a scrambled series of ink-marks. In such a case it would be fanciful to think that any arbitrary foliation is just as good as that which treats the block as a stack of pages (Fig. 2.2). Analogously, the evolving geometry of space over time in Friedman models discloses the natural foliation of space-time in such a universe. The evolving, dynamic geometry of space, like the cartoon figure, would be destroyed by just any arbitrary slicing up of space-time (Fig. 2.3). In a Friedman universe, then, there is a preferred slicing of space-time along a cosmic time parameter in line with certain natural symmetries.

Now as a parameter, cosmic time measures the duration of the universe as a whole in an observer-independent way; that is to say, the lapse of cosmic

---

[50] Willem de Sitter, "On the Relativity of Inertia," in *Koninglijke Nederlandse Akademie van Wetenschappen Afdeling Wis. en Natuurkundige Wetenschappen, Proceedings of the Section of Science* 19 (1917): 1217-1225.

[51] A. Friedman, "Über die Krümmung des Raumes," *Zeitschrift für Physik* 10 (1922): 377-386.

[52] See Charles W. Misner, Kip S. Thorne, and John Archibald Wheeler, *Gravitation* (San Francisco: W. H. Freeman, 1973), 713-714; Kanitscheider, *Kosmologie*, 182-197.

Divine Timelessness

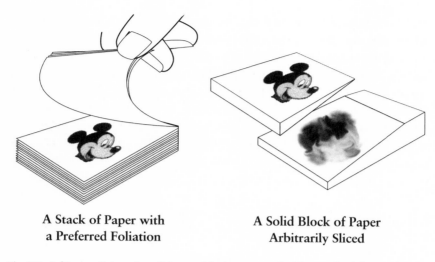

A Stack of Paper with a Preferred Foliation

A Solid Block of Paper Arbitrarily Sliced

Fig. 2.2: Arbitrary slicing of a solid paper block contrasted with a stack of sheets of paper.

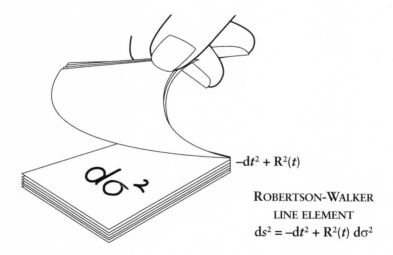

$-dt^2 + R^2(t)$

ROBERTSON-WALKER LINE ELEMENT
$$ds^2 = -dt^2 + R^2(t)\, d\sigma^2$$

Fig. 2.3: Preferred slicing of space-time as disclosed by the natural geometrical symmetries in the Friedman model. The geometry of space ($d\sigma^2$) evolves over time ($-dt + R^2[t]$).

time is the same for all observers. Nevertheless, cosmic time is related to the local times of a special group of observers called "fundamental observers." These are hypothetical observers who are at rest with respect to the expansion of space itself. It is important to realize that despite potentially misleading expressions such as "the Big Bang," the expansion of the universe should not be thought of as the explosion of material into a previously existing empty space. Rather in Friedman models it is space itself which is expanding, and the galaxies are fixed in space and are simply "riding along" with the expanding space. The easiest way to envision this is to imagine a balloon with buttons glued to its surface. As you blow up the balloon, the buttons, though stuck in place, move away from each other because the balloon itself is expanding. The surface of the balloon is the two-dimensional analogue to space, and the buttons are like the galaxies. As space expands, the galaxies recede from each other, even though they are fixed in space. Now fundamental observers are hypothetical observers associated with the galaxies. As time goes on and the expansion of space proceeds, each fundamental observer remains in the same place, though his spatial separation from fellow fundamental observers increases. Cosmic time relates to these observers in that their local times all coincide with cosmic time in their vicinity. Because of their mutual recession, the class of fundamental observers do not serve to define a global inertial frame, technically speaking, even though all of them are at rest. But since each fundamental observer is at rest with respect to space, the events which he calculates to be simultaneous will coincide locally with the events which are simultaneous in cosmic time. What this implies is that, contrary to premise (2), it does not follow from the correctness of STR that if God is in time, then He is in the time of one or more inertial frames.[53] For if God exists in cosmic time, there is no universal inertial frame with which He can be associated.

Thus, on a cosmic scale, we seem to have that universality of time and absolute simultaneity of events which the Special Theory had denied. G. J. Whitrow of London's Imperial College of Science and Technology asserts, "in a universe that is characterized by the existence of a cosmic time, relativity is reduced to a local phenomenon, since this time is world-wide and independent of the observer."[54] Based on a cosmological rather than a local perspective, cosmic time serves to restore to us our intuitive notions of universal time and absolute simultaneity which STR denied.

The question, then, becomes an empirical one: Does cosmic time exist? Do we live in a Friedman universe? The evidence strongly suggests that we do. According to the British Royal Astronomer Martin Rees, "The most remarkable outcome of 50 years of observational cosmology has been the realization that the universe is more isotropic and uniform than the pioneer

---

[53] In the sense that God exists in the time of the inertial frame of each fundamental observer, there is no objection, since all their local times fuse into one cosmic time.
[54] Whitrow, *Natural Philosophy of Time*, 371; cf. 302.

theorists of the 1920's would ever have suspected."[55] The recent findings of the COBE satellite, which has measured the uniformity of the cosmic background radiation to one part in 100,000, have dramatically underscored this conclusion. "We have strong evidence that the universe as a whole is predominantly homogeneous and isotropic," states Whitrow, "and this conclusion . . . is a strong argument for the existence of cosmic time."[56] Thus, far from "taking away from space and time the last remnant of physical objectivity," as Einstein thought at first, GTR through its cosmological applications appears to give back what STR had removed.

The defender of divine temporality may accordingly hold that God exists in cosmic time. Already in 1920, on the basis of Einstein's and de Sitter's cosmological models, Eddington hinted at a theological interpretation of cosmic time:

> In the first place, absolute space and time are restored for phenomena on a cosmical scale. . . . The world taken as a whole has one direction in which it is not curved; that direction gives a kind of absolute time distinct from space. Relativity is reduced to a local phenomenon; and although this is quite sufficient for the theory hitherto described, we are inclined to look on the limitation rather grudgingly. But we have already urged that the relativity theory is not concerned to deny the possibility of an absolute time, but to deny that it is concerned in any experimental knowledge yet found; and it need not perturb us if the conception of absolute time turns up in a new form in a theory of phenomena on a cosmical scale, as to which no experimental knowledge is yet available. Just as each limited observer has his own particular separation of space and time, so a being co-extensive with the world might well have a special separation of space and time natural to him. It is the time for this being that is here dignified by the title "absolute."[57]

Notice that Eddington is quite willing to call cosmic time "absolute" in view of its independence from space, that is to say, its status as a parameter. Relativistic time is only a local time, whereas cosmic time, being non-local, is the true time. Although in 1920 there was no empirical evidence for cosmic time, within a few short years astronomical evidence confirmed the prediction of the Friedman model of a universal expansion and, hence, of cosmic

---

[55] Martin J. Rees, "The Size and Shape of the Universe," in *Some Strangeness in the Proportion,* ed. Harry Woolf (Reading, Mass.: Addison-Wesley, 1980), 293. *Isotropy* is the property of being the same in all directions.
[56] Whitrow, *Natural Philosophy of Time,* 307.
[57] Arthur Eddington, *Space, Time and Gravitation,* Cambridge Science Classics (Cambridge: Cambridge University Press, 1920; rep. ed.: 1987), 168.

time. This cosmic time would, says Eddington, be the time of an omnipresent being. Cosmic time is not merely the "fusion" of all the local times recorded by the separate fundamental observers, but, even *more* fundamentally, it is the time which measures the duration of the universe. As the measure of the proper time of the universe, cosmic time also measures the duration of and lapse of time for a temporal being coextensive with the world. For Eddington, it is the time of this being that deserves to be called "absolute."

Such an affirmation will be typically met with passionate disclamations. Any equivalence of cosmic time with Newton's absolute time is usually vigorously repudiated by relativistic scientists. But here one must not confuse the various senses of "absolute." Eddington is not claiming that cosmic time is metaphysically necessary or independent of physical measures. Rather he is saying that cosmic time is not tied to inertial frames and so is privileged. One of the most intriguing indications that cosmic time does represent the physical equivalent of Newton's absolute time is the surprising demonstration by E. A. Milne and W. H. McCrea that all the results of GTR-based Friedman cosmology can be recovered by Newtonian physics and in a way that is simpler than Einstein's cumbersome mathematics! Milne and McCrea were able to reproduce all the results of Big Bang cosmology by means of a material universe expanding in empty, classical space through classical time.[58] Comparing relativistic and Newtonian cosmology, Kerszberg observes, "as far as the prediction of the overall history of the universe is concerned, the equivalence seems to be total."[59] This implies, in Bondi's words, that GTR "cannot be expected to explain any major features in any different or better way than Newtonian theory."[60] In particular the concept of cosmic time in GTR-based models corresponds to absolute time in the Newtonian model. Schücking points out that the main asset of the Milne-McCrea formulation was that it gave exactly the same equations for the time development of the universe as the Friedman theory and yet allowed a much simpler derivation.[61] This is not to suggest that Newtonian theory is correct after all; we have already seen how Lorentz was forced to modify Newtonian physics on the local level. But the equivalence of Milne-McCrea Newtonian cosmology with GTR-based Friedman cosmology is a convincing demonstration that cosmic time is,

---

[58] E. A. Milne, *Relativity, Gravitation and World Structure* (Oxford: Clarendon, 1935); idem, "A Newtonian Expanding Universe," *Quarterly Journal of Mathematics* 5 (1934): 64-72; W. H. McCrea, "On the Significance of Newtonian Cosmology," *Astronomical Journal* 60 (1955): 271-274.

[59] Pierre Kerszberg, "On the Alleged Equivalence between Newtonian and Relativistic Cosmology," *British Journal for the Philosophy of Science* 38 (1987): 349.

[60] Hermann Bondi, *Cosmology* (Cambridge: Cambridge University Press, 1952), 70-71.

[61] E. L. Schücking, "Newtonian Cosmology," *Texas Quarterly* 10 (1967): 274.

indeed, the physical equivalent of Newtonian absolute time. Thus, Bondi likens cosmic time with Newton's uniform, omnipresent, and even-flowing time, which enables all observers to synchronize their clocks to a single time.[62] Kerszberg concludes, "On the whole, the equivalence between Newtonian and relativistic cosmology only reinforces the conviction that cosmic time is indeed a necessary ingredient in the formalisation of a relativistic cosmology, however alien to general relativity and congenial to Newton's theory the notion of universal synchronisation might seem."[63]

Now at this point the advocate of divine timelessness may think that he has just been inadvertently delivered the trump card. For cosmic time had a beginning; the Big Bang event represents not just the origin of all the matter and energy in the universe, but the origin of space-time itself. There is no moment "before" the Big Bang, since time originated at the Big Bang. So if God is temporal and time had a beginning, God must have had a beginning. But obviously, God did not come into existence with the Big Bang, or at any other time, for His existence is beginningless and endless. God must therefore transcend time and is thus timeless.

The Newtonian will be unfazed by this objection, however, for he may plausibly construe cosmic time as but an empirical measure of God's time since the moment of creation. Newton himself implies such an interpretation when he writes:

> Absolute time, in astronomy, is distinguished from relative [time], by the equation or correction of the apparent time. For the natural days are truly unequal, though they are commonly considered as equal, and used for a measure of time; astronomers correct this inequality that they may measure the celestial motions by a more accurate time. It may be, that there is no such thing as an equable motion, whereby time may be accurately measured. All motions may be accelerated and retarded, but the flowing of absolute time is not liable to any change. The duration or perseverance of the existence of things remains the same, whether the motions are swift or slow, or none at all: and therefore this duration ought to be distinguished from what are only sensible measures thereof; and from which we deduce it, by means of the astronomical equation.[64]

Cosmic time provides an approximate measure of God's absolute time and of His co-existence with the universe since the moment of creation. While this

---

[62] Bondi, *Cosmology*, 70-71.
[63] Kerszberg, "Equivalence," 376.
[64] Newton, *Principles of Natural Philosophy*, 1:7-8.

empirical measure of time had a beginning in the Big Bang, time itself did not. Thus God existed literally before the Big Bang event in absolute time. Newton believed that the "flow" of absolute time would exist even in the utter absence of events—as he says, "whether the motions are swift or slow, or none at all." For time, on His view, is the immediate effect of God's merely existing. Thus, even if there were no events prior to creation, time would still exist as the duration of God's being.

It ought to be noted, however, that the view that time existed prior to creation does not depend upon adopting Newton's belief that time can exist in the absence of events. The seventeenth-century German polymath Gottfried Wilhelm Leibniz opposed Newton in this regard, maintaining that time is a relation between events and so could not exist if there were no events. Leibniz therefore held that time began at the moment of creation with the occurrence of the first event. But even granted a Leibnizian relational view of time, it can still make sense to talk about time prior to creation. For the events which serve to generate time need not be physical events; a sequence of mental events would suffice. Suppose, for example, that God were to be counting down toward the moment of creation: " . . . three . . . two . . . one . . . Let there be light!" In such a case the mental events of counting would generate a temporal succession of moments. Or God could have created angelic beings prior to the Big Bang which were undergoing a succession of mental states. Thus, whether one adopts a Newtonian (substantival) or Leibnizian (relational) view of time, it makes sense to talk about time prior to the inception of physical, cosmic time, which is but an empirical measure of time itself. Indeed, I take the coherence of this thought experiment to be a knock-down argument that STR, or any other scientific theory, does not furnish a correct description of time itself. At best, scientific accounts describe our *measures* of time, but not time *itself*.

In conclusion, Relativity Theory does not provide good grounds for thinking that God is timeless. The Einsteinian interpretation of STR is based essentially upon an untenable and obsolete verificationist epistemology and so cannot force abandonment of the classical concept of time. Moreover, GTR in its cosmological application furnishes us with a cosmic time parameter which may be plausibly interpreted as the appropriate measure of God's time since the moment of creation. The past finitude of cosmic time does not imply the finitude of God's time, for whether one adopts a substantival or a relational view of time, it is coherent to speak of God's existing temporally prior to the creation of the universe and the beginning of cosmic time.

III. The Incompleteness of Temporal Life

EXPOSITION

An important argument in favor of divine timelessness rests on the claim that the fleeting nature of temporal life is incompatible with the life of a most perfect being such as God is. For example, in his study of time and eternity, the Fordham University philosopher Brian Leftow draws upon Boethius's characterization of eternity as *complete possession all at once of interminable life* in order to argue for the defectiveness of temporal existence.[65] Leftow points out that a temporal being is unable to enjoy what is past or future for it. The past is gone forever, and the future is yet to come. The passage of time thus renders it impossible for any temporal being to possess all its life at once. Even God, if He is temporal, cannot reclaim the past. Leftow emphasizes that even perfect memory cannot substitute for reality: "the past itself is *lost,* and no memory, however complete, can take its place—for confirmation, ask a widower if his grief would be abated were his memory of his wife enhanced in vividness and detail."[66] By contrast, a timeless God lives all His life at once because He literally has no past or future and so suffers no loss. Therefore, since God is the most perfect being, He is timeless.

We can formulate this argument as follows:

1. God is the most perfect being.

2. The most perfect being has the most perfect mode of existence.

3. Temporal existence is a less perfect mode of existence than timeless existence.

4. Therefore, God has the most perfect mode of existence.

5. Therefore, God has a timeless mode of existence.

CRITIQUE

Here I think we have an argument for divine timelessness which is really promising. The premises of the argument rest on very powerful intuitions about the irretrievable loss that arises through the experience of temporal pas-

---
[65] Brian Leftow, *Time and Eternity,* Cornell Studies in Philosophy of Religion (Ithaca, N.Y.: Cornell University Press, 1991), 278.
[66] Ibid.

sage, a loss which intuitively should not characterize the experience of a most perfect being. The fleeting nature of temporal life was brought home to me unexpectedly and powerfully as I read aloud to our children Laura Ingalls Wilder's account of life in the American Midwest during the late 1800s in her *Little House in the Big Woods*. Here are the final paragraphs of that book:

> The long winter evenings of firelight and music had come again. . . . Pa's strong, sweet voice was softly singing:
>
> "Shall auld acquaintance be forgot,
> And never brought to mind?
> Shall auld acquaintance be forgot,
> And the days of auld lang syne?
> And the days of auld lang syne, my friend,
> And the days of auld lang syne,
> Shall auld acquaintance be forgot,
> And the days of auld lang syne?"
>
> When the fiddle had stopped singing Laura called out softly, "What are days of auld lang syne, Pa?"
>
> "They are the days of a long time ago, Laura," Pa said. "Go to sleep, now."
>
> But Laura lay awake a little while, listening to Pa's fiddle softly playing and to the lonely sound of the wind in the Big Woods. She looked at Pa sitting on the bench by the hearth, the firelight gleaming on his brown hair and beard and glistening on the honey-brown fiddle. She looked at Ma, gently rocking and knitting.
>
> She thought to herself, "This is now."
>
> She was glad that the cosy house, and Pa and Ma and the firelight and the music, were now. They could not be forgotten, she thought, because now is now. It can never be a long time ago.[67]

What makes this passage so poignant is that as we read it today we realize that the time which for Laura Ingalls was so real, which was "now," is no longer now, but is gone forever. Pa and Ma are gone, the American frontier they struggled to win is gone, the years Laura Ingalls called "those happy golden days" are all gone, gone forever, never to be reclaimed. Time has a savage way of gnawing away at life, leaving it transitory and incomplete, so that life in its fullness can never be enjoyed by any temporal being.

---

[67] Laura Ingalls Wilder, *Little House in the Big Woods* (New York: Harper & Row, 1932), 237-238.

The force of these considerations is such that Stump and Kretzmann, whose 1981 article "Eternity" in the *Journal of Philosophy* sparked a revival of interest in the doctrine of divine timelessness, have rested their case for God's atemporality solely on the shoulders of this argument. They comment,

> No life . . . that is imperfect in its being possessed with the radical incompleteness entailed by temporal existence could be the mode of existence of an absolutely perfect being. A perfectly possessed life must be devoid of any past, which would be no longer possessed, and of any future, which would be not yet possessed. The existence of an absolutely perfect being must be an indivisibly persistent present actuality.[68]

Their claim that the life of a most perfect being must be an indivisible actuality has, I think, a good deal of plausibility.

Some philosophers of time might try to avert the force of this argument by adopting a view of time—of which we shall have more to say later—according to which things and events do not in fact come to be or pass away. According to this view of time, often called the "tenseless" or "static" view of time, the past and future are just as real as the present. The difference between past, present, and future is usually explained as just a subjective illusion of human consciousness. For the people located in 1868, for example, the events of 1868 are present, and we are future; by the same token, for the people living in 2050 it is the events of 2050 that are present, and we are past. Time is akin to a spatial line, and all the points of the line are equally real. On such a view of time, if something has a finite lifetime, it does not come into being at a certain point and go out of being at a later point. Rather it just exists at those two points and all the points in between. The longer a thing's temporal extension, the longer its lifetime. If the temporal extensions of two persons overlap, then they will regard themselves as both present during that period of overlap. If one has a longer time line than the other, then the person with the longer time line will regard the other as at some point no longer present; but, say the philosophers who hold this view, if that person is philosophically informed, he will not regard his fellow as non-existent. Albert Einstein, who came to adopt this view of time, took this idea so seriously that when his lifelong friend Michael Besso died, he tried to comfort Besso's surviving son and sister by writing, "This signifies nothing. For us believing physicists the distinction between past, present, and future is only an illusion,

---

[68] Eleonore Stump and Norman Kretzmann, "Prophecy, Past Truth, and Eternity," *Philosophical Perspectives* 5 (1991): 395; cf. idem, "Eternity, Awareness, and Action," *Faith and Philosophy* 9 (1992): 463.

even if a stubborn one."⁶⁹ On this view of time no temporal being ever really loses its past or has not yet acquired its future. Just as things are extended in space, so they are also extended in time. A temporal being has nothing to lose and nothing to gain; it just exists tenselessly at its temporal locations. Thus, a temporal God would exist at all temporal locations without beginning or end to His temporal extension. On this view of time God does not lose or acquire portions of His life.

The problem with this escape route is that it fails to appreciate that the argument is based on the *experience* of temporal passage, rather than on the objective reality of temporal passage itself. The flow of time is an ineradicable part of the experience of a temporal being. Even if the future never becomes and the past is never really lost, the fact remains that for a temporal being the past is lost *to him* and the future is not accessible *to him*. As H. G. Wells's celebrated Time Traveler, who believed that time was a fourth dimension of space, remarked, "Our mental existences, which are immaterial and have no dimensions, are passing along the Time-Dimension with a uniform velocity from the cradle to the grave."⁷⁰ Even if the cradle and the grave are just as real as the present, we still find ourselves experientially at some point in between, and events which are located at times earlier than that point are irretrievably lost to us, and events later than that point can only be anticipated. For this reason a tenseless or static theory of time does nothing to alleviate the loss occasioned by our experience of temporal becoming. I dare say that the bereaved find little comfort in the thought that a deceased loved one exists tenselessly at earlier temporal coordinates than those which they occupy. Time's tooth gnaws away at our experience of life regardless of the tenseless existence of all events making up one's life. For this reason, it would be futile to attempt to elude the force of this argument by postulating a temporal deity in a tenseless time.

Other philosophers, observing that this argument concerns, not temporal passage itself, but our *experience* of temporal passage, have suggested a different way around the argument. The fleetingness of our experience derives from the limits of what psychologists call our "specious present," that is to say, our subjective awareness of what is "now." The average person's now-awareness is just a fraction of a second. But the longer one's specious present, the less fleeting one's experience of life would be. If we could imagine some-

---

⁶⁹ Letter of Albert Einstein, March 21, 1955, cited in *Albert Einstein: Creator and Rebel*, Banesh Hoffmann with Helen Dukas (London: Hart-Davis, MacGibbon, 1972), 258.
⁷⁰ H. G. Wells, *The Time Machine* (New York: Berkeley, 1957), 10. Of course, the "passing along" must have reference to our *experience* of time's flow; contrary to Wells, psychological time passes at various rates.

one who experienced a specious present which had the same duration as his entire life, such a person would experience his life all at once. These considerations have led William Alston, a noted Christian philosopher of the University of Syracuse, to claim that God's specious present has the same temporal extension as the whole of time, so that God has, indeed, at least experientially, complete possession all at once of interminable life. He writes,

> Just expand the specious present to cover all of time, and you have a model for God's awareness of the world. . . . a being with an infinite specious present would not, so far as his awareness is concerned, be subject to temporal succession at all. There would be no further awareness to succeed the awareness in question. *Everything* would be grasped in one temporally unextended awareness.[71]

Such a model would enable us to hold to God's being temporal and yet experiencing His entire life at once as a whole.

Nevertheless, a little reflection reveals that this model exacts far too high a price for the benefits it offers. (i) The reason we human beings have a specious present is due to our physical limitations, particularly the finite velocity of the transmission of signals along our nervous system. Because we do not have instantaneous transmission of such signals, there is a minimum threshold of the psychological present, so that events which occur too quickly cannot be experienced by us as consecutively present. But God, as unembodied Mind possessing maximal cognitive excellence, should possess no minimal, finite psychological present at all, much less an infinitely extended one. He is not dependent upon finite velocity neural processes which would slow down His apprehension of present events. And being maximally excellent cognitively, we should rather expect that He be able, rather than unable, to distinguish discrete, consecutive events as present. As one commentator has remarked, a God with an everlasting specious present would be infinitely slow on the uptake![72] In a literal sense, He would be mentally retarded. (ii) The specious present gathers into one now-awareness a period of time up to the present moment. Thus, if God had a specious present covering the whole of

---

[71] William P. Alston, "Hartshorne and Aquinas: A Via Media," in *Existence and Actuality*, ed. John B. Cobb, Jr. and Franklin I. Gamwell (Chicago: University of Chicago Press, 1984), 91. In all fairness to Alston, it must be admitted that he is using the specious present as "an intelligible model for a *nontemporal* knowledge of a temporal world" (p. 90, emphasis added). For a literal affirmation of God's having an everlasting specious present, see Grace M. Jantzen, *God's World, God's Body,* with a foreword by John MacQuarrie (London: Darton, Longman, and Todd, 1984), 65.

[72] Fitzgerald, "Relativity Physics and the God of Process Philosophy," 267. Fitzgerald goes on to say, "This makes God out to be a sort of infinitely sluggish observer of the passing scene. . . . Contrary to what appears at first, it is a defect rather than a merit to have a specious present which is all inclusive."

time, He would not experience His specious present until He had endured to the end of time. But then, although God at that instant becomes aware of the succession of all events, it is too late for Him to do anything about them, for they are already past by that point. Thus, God could not respond to individual events in time. God's providence is therefore obliterated by such a model. Worse, God could not even know what He Himself had done throughout history until it was over. How He could even act in history without any consciousness of what was happening at the time the events occurred remains a mystery. A sort of backward causation would seem to be necessary to explain God's acts in time. All these untoward consequences result if time in fact has an end. But if time has no end, as the Christian doctrine of the afterlife teaches, then God *never* becomes conscious. There is no point at which all His cognitions of individual events can be gathered into a specious present, since there will always be time after that. Thus, the model becomes self-contradictory, for in order to have a specious present which takes in all of unending time, God's becoming conscious is indefinitely postponed such that He never has a specious present. (iii) It might be suggested that we loose the model from its physical basis in neurology and interpret God's specious present merely on the *analogy* of our specious present. God just has at every point in time a specious present which takes in the whole of time. But as recent studies in the philosophy of language have shown, the ability to apprehend tenses, that is to say, the ability to know what is happening *now,* is essential to timely action. If God has the same specious present at every moment of time, then He has neither memory nor foreknowledge nor changing now-awarenesses. Thus, He is rendered utterly impotent to act in a timely fashion, since He never knows what time it is. Instead of a variety of now-awarenesses at different times, He has at each time the same now-awareness. Hence, He is incapacitated to effect something at the time at which He is located. In short, it seems to me that the theory of God's having an everlasting specious present is utterly inept and so affords no escape from the present argument.

Perhaps, however, the realization that the argument for divine timelessness from the incompleteness of temporal life is essentially experiential in character opens the door for a temporalist alternative. When we recall that God is perfectly omniscient and so forgets absolutely nothing of the past and knows everything about the future, then time's tooth is considerably dulled for Him. His past experiences do not fade as ours do, and He has perfect recall of what He has undergone. To be sure, the past itself is gone (given a tensed or dynamic view of time), but His experience of the past remains as vivid as ever. A fatal flaw in Leftow's analysis is his assumption that God, like the widower, has actually lost the persons He loves and remembers. But according to Christian theism, this assumption is false. Those who perish physically live on in the afterlife, where they continue to be real and present to God. At worst, what are past are the experiences God has enjoyed of those persons, for example, Jones's coming to faith in Christ. But in the afterlife Jones lives on with God, and God can recall as though it were present His experience of Jones's conversion. So it is far from obvious that the experience of temporal passage is so melancholy an affair for an omniscient God as it is for us.

Moreover, it needs to be kept in mind that the life of a perfect person may have to be characterized by the incompleteness which would in other contexts be considered an imperfection. There is some evidence that consciousness of time's flow can actually be an enriching experience.[73] R. W. Hepburn cautions against downplaying the importance of the flow of consciousness in awareness of music, for example. Music appreciation is not merely a matter of apprehending tenselessly the succession of sounds. Quoting Charles Rosen to the effect that, "The movement from past to future is more significant in music than the movement from left to right in a picture," Hepburn believes that the phenomenon of music calls into question any claim that a perfect mode of consciousness would be exclusively atemporal. All this goes to call into question premise

> 3. Temporal existence is a less perfect mode of existence than timeless existence

of the argument for divine timelessness from the incompleteness of temporal life. Timeless life may not be the most perfect mode of existence of a perfect person.

Still, I think that we must admit that the argument has some force and could motivate justifiably a doctrine of divine timelessness in the absence of countermanding arguments. The question then will be whether the reasons for affirming divine temporality do not overwhelm this argument for divine timelessness.

Conclusion

In summary, we have seen that the arguments for divine timelessness are inconclusive. While God's timelessness does follow from divine simplicity or

---

[73] See the very interesting piece by R. W. Hepburn, "Time-Transcendence and Some Related Phenomena in the Arts," in *Contemporary British Philosophy*, 4th series, ed. H. D. Lewis, Muirhead Library of Philosophy (London: George Allen & Unwin, 1976), 152-173.

immutability, those doctrines are even more controversial than the doctrine of divine timelessness and so furnish no grounds for adoption of the view that God is timeless. The appeal to the Special Theory of Relativity in order to ground belief in God's atemporality is unpersuasive, since the defender of God's temporality may justifiably challenge the verificationist epistemological underpinnings of the theory and so distinguish between time itself and our physical measures thereof. Our inability to detect empirically relations of absolute simultaneity is no reason to think that such relations do not exist. Indeed, such relations may be plausibly grounded in a preferred reference frame associated with God's "now" in absolute time. Finally, the argument based on the incompleteness of temporal life is essentially an experiential argument, whose force is mitigated in the case of God. Still, this last argument does have some force and so needs to be weighed against whatever arguments can be offered on behalf of divine temporality.

Recommended Reading

Divine Simplicity and Immutability

Hughes, Christopher. *On a Complex Theory of a Simple God.* Cornell Studies in the Philosophy of Religion. Ithaca, N.Y.: Cornell University Press, 1989.

Morris, Thomas V. *Anselmian Explorations,* pp. 98-123. Notre Dame, Ind.: University of Notre Dame Press, 1987.

Relativity Theory

Newton, Isaac. *Sir Isaac Newton's "Mathematical Principles of Natural Philosophy" and His "System of the World,"* 2 vols., *Scholium* to the Definitions (1:6-11) and *General Scholium* (2:544-546). Translated by Andrew Motte. Revised with an appendix by Florian Cajori. Los Angeles: University of California Press, 1966.

_____. "On the Gravity and Equilibrium of Fluids." In *Unpublished Scientific Papers of Isaac Newton.* Edited by A. Rupert Hall and Marie Boas Hall. Cambridge: Cambridge University Press, 1962.

McGuire, J. E. "Newton on Place, Time, and God: An Unpublished Source." *British Journal for the History of Science* 11 (1978): 114-129.

Einstein, Albert. "On the Electrodynamics of Moving Bodies." Translated by Arthur Miller. Appendix to Arthur I. Miller, *Albert Einstein's Special Theory of Relativity.* Reading, Mass.: Addison-Wesley, 1981.

_____. *Relativity: The Special and General Theories.* London: Methuen, 1954.

Miller, Arthur I. *Albert Einstein's Special Theory of Relativity.* Reading, Mass.: Addison-Wesley, 1981.

Taylor, J. G. *Special Relativity.* Oxford Physics Series. Oxford: Clarendon, 1975.

Eddington, Arthur. *Space, Time and Gravitation.* Cambridge Science Classics. Cambridge: Cambridge University Press, 1920; rep. ed.: 1987.

Norton, John D. "Philosophy of Space and Time." In *Introduction to the Philosophy of Science*, pp. 3-56. Edited by Merilee Salmon. New Jersey: Prentice-Hall, 1992.

Holton, Gerald. *Thematic Origins of Scientific Thought: Kepler to Einstein*. Cambridge, Mass.: Harvard University Press, 1973.

_____. "Mach, Einstein and the Search for Reality." In *Ernst Mach: Physicist and Philosopher*, pp. 165-199. Boston Studies in the Philosophy of Science 6. Dordrecht: D. Reidel, 1970.

_____. "Where Is Reality? The Answers of Einstein." In *Science and Synthesis*, pp. 45-69. Edited by UNESCO. Berlin: Springer-Verlag, 1971.

Zahar, Elie. "Why Did Einstein's Programme Supersede Lorentz'?" *British Journal for the Philosophy of Science* 24 (1973): 95-123, 223-262.

Illy, Jozsef. "Einstein Teaches Lorentz, Lorentz Teaches Einstein. Their Collaboration in General Relativity, 1913–1920." *Archive for History of Exact Sciences* 39 (1989): 247-289.

Sklar, Lawrence. "Time, Reality, and Relativity." In *Reduction, Time and Reality*, pp. 129-142. Edited by Richard Healey. Cambridge: Cambridge University Press, 1981.

Fitzgerald, Paul. "Relativity Physics and the God of Process Philosophy." *Process Studies* 2 (1972): 251-276.

Misner, Charles W., Kip S. Thorne, and John Archibald Wheeler. *Gravitation*. San Francisco: W. H. Freeman, 1973.

A Companion to Philosophy of Religion. Edited by Philip L. Quinn and Charles Taliaferro. Blackwell Companions to Philosophy. Oxford: Blackwell, 1997. S.v. "Theism and Physical Cosmology," by William Lane Craig.

The Incompleteness of Temporal Life

Leftow, Brian. *Time and Eternity*, pp. 278-279. Cornell Studies in the Philosophy of Religion. Ithaca, N.Y.: Cornell University Press, 1991.

Gale, Richard M. "The Static vs. the Dynamic Temporal: Introduction." In *The Philosophy of Time: A Collection of Essays*, pp. 169-178. Edited by Richard M. Gale. New Jersey: Humanities Press, 1968.

Craig, William Lane. "On the Argument for Divine Timelessness from the Incompleteness of Temporal Life." *Heythrop Journal* 38 (1997): 165-171.

3

DIVINE TEMPORALITY

THOMAS AQUINAS CLAIMED that God is timeless and so sees all of time from beginning to end, just as a man on a watchtower sees the whole stretch of a caravan passing by on the road below. Thus, the whole of time is present to eternity. Reacting to Thomas's claim, the medieval Scottish theologian John Duns Scotus protested,

> Eternity will not, by reason of its infinity, be present to any *non-existent* time.... If (assuming the impossible) the whole of time were *simultaneously* existent, the whole would be simultaneously present to eternity.... For the "now" of eternity is formally infinite and therefore formally exceeds the "now" of time. Nevertheless it does not co-exist with *another* "now".[1]

On Scotus's understanding of time and eternity, God co-exists only with the present moment or "now." He is eternal in the sense that He endures forever.

Again, we want to ask what reasons might be given for adopting this temporalist understanding of divine eternity. Of the various arguments on behalf of divine temporality, three stand out as especially significant.

I. The Impossibility of Atemporal Personhood

EXPOSITION

We have seen that Isaac Newton founded his belief in the existence of absolute time on God's infinite temporal duration. But so far as I can tell, Newton never offered any argument for thinking God to be temporal—he just asserted it. He regarded temporality and spatiality as inherent dispositions of being; that is to say, anything that exists must exist in time and space. But this assumption is far from obvious. Indeed, quite the contrary, it seems easy to

---
[1] John Duns Scotus, *Ordinatio* 1. 38-39. 9-10.

conceive of God as transcending space, since He is incorporeal. Moreover, philosophers often regard abstract entities such as numbers or sets as existing in neither time nor space. So why could God not exist timelessly? Is there no logically conceivable world in which God exists and time does not?

According to the Christian doctrine of creation, God's decision to create a universe was a freely willed decision from which God could have refrained. We can conceive, then, of a possible world in which God does refrain from creation, a world which is empty except for God. Would time exist in such a world? Certainly it would if God were changing, experiencing a stream of consciousness. As we have seen, even a succession in the contents of consciousness is sufficient to generate a temporal series.

But suppose God were altogether changeless. Suppose that He did not experience a succession of thoughts but grasped all truth in a single, changeless intuition. Would time exist? A relationalist like Leibniz would say no, for there are no events to generate a relation of *earlier than* or *later than*. There is just a single, timeless state.

It is true that in recent years there has been a good deal written about the possibility of time without change, and most contemporary relationalists espouse a view which allows there to be changeless periods of time sandwiched in between periods of change.[2] But I know of no relationalist account that would allow a totally changeless world such as we are envisioning to be temporal. Such a world would, indeed, seem to be just a single, timeless state.

Newton would have disagreed, of course. For him timeless existence was a logical impossibility. But my point is that no reason has been offered why we should side with Newton on this score rather than with Leibniz, whose view seems extremely plausible.

If timeless existence as such is not demonstrably impossible, then, why should we think that God could not exist timelessly? Let us stick with our envisioned empty world in which God alone exists. Why could God not exist timelessly in such a world?

"Because God is personal!" is the answer given by certain advocates of divine temporality. They contend that the idea of a timeless person is incoherent and therefore God must be temporal. They argue that in order to be a person, one must possess certain properties which inherently involve time. Since God is essentially personal, He therefore cannot be timeless.

---

[2] See the seminal paper by Sidney Shoemaker, "Time without Change," *Journal of Philosophy* 66 (1969): 363-381.

We can formulate this argument as follows (using x, y, z to represent certain properties to be specified later):

1. Necessarily, if God is timeless, He does not have the properties x, y, z.

2. Necessarily, if God does not have the properties x, y, z, then God is not personal.

3. Necessarily, God is personal.

4. Therefore, necessarily, God is not timeless.

The argument, if successful, shows that timelessness and personhood are incompatible and, since God is essentially personal, it is timelessness which must be jettisoned.

CRITIQUE

The defender of divine timelessness may attempt to turn back this argument either by challenging the claim that the properties in question are necessary conditions of personhood or by showing that a timeless God could possess the relevant properties after all. So what are the properties x, y, z that the advocate of divine temporality is talking about?

In his article "Conditions of Personhood,"[3] Daniel Dennett, a philosopher who specializes in the philosophy of mind, delineates six different conceptions of personhood, each of which lays down a necessary condition of any individual P's being a person:

P is a person only if:

i. P is a rational being.
ii. P is a being to which states of consciousness can be attributed.
iii. Others regard (or can regard) P as a being to which states of consciousness can be attributed.

---

[3] Daniel Dennett, "Conditions of Personhood," in *The Identities of Persons*, ed. Amelie Oksenberg Rorty (Berkeley: University of California Press, 1976), 175-196. Dennett's criteria were first used in defense of divine, timeless personhood by William E. Mann, "Simplicity and Immutability in God," *International Philosophical Quarterly* 23 (1983): 267-276.

iv. *P* is capable of regarding others as beings to which states of consciousness can be attributed.
v. *P* is capable of verbal communication.
vi. *P* is self-conscious; that is, *P* is capable of regarding him/her/itself as a subject of states of consciousness.

All of these criteria depend in some way on *P*'s having or being said to have consciousness. So, as an initial step in assessing the present argument, we may ask whether the concept of a conscious, timeless being is possible.

John Lucas is one of those philosophers who maintains that this is not possible. He writes,

> Time is not a thing that God might or might not create, but a category, a necessary concomitant of the existence of a personal being, though not of a mathematical entity. This is not to say that time is an independent category, existing independently of God. It exists because of God: not because of some act of will on His part, but because of His nature: if ultimate reality is personal, then it follows that time must exist. God did not make time, but time stems from God.[4]

On Lucas's view, even in an otherwise empty world, time would exist if a personal God exists. Unfortunately, Lucas never explains why personal consciousness could not be unchanging and therefore, plausibly, atemporal. Why could not the contents of God's consciousness in such a world be comprised exclusively of such changelessly true beliefs as "No human beings exist," "7+5=12," "Anything that has a shape has a size," "If I were to create a world of free creatures, they would fall into sin," and so forth? If God never acquires any new beliefs and never loses any beliefs, why could not such a changeless consciousness of truth be plausibly regarded as timeless? Why think that such a changeless, timeless consciousness is impossible? Here Lucas has nothing to say. He confesses, "My claim . . . that time is a concomitant of consciousness, is of course only a claim, and I have been unable to argue for it, except by citing poetry. . . . arguments would be better."[5]

Indeed, they would! So what arguments are there against the possibility of an atemporal consciousness? Richard Gale, a well-known philosopher of time, would make short work of the question: "the quickest and most direct

---
[4] J. R. Lucas, *The Future: An Essay on God, Temporality, and Truth* (Oxford: Basil Blackwell, 1989), 213; cf. 212.
[5] Ibid., 175.

way of showing the absurdity of a timeless mind is as follows: A mind is conscious, and consciousness is a temporally elongated process."[6] The difficulty with Gale's reasoning, however, is that he fails to show that being temporally extended is an *essential* property of consciousness, rather than just a *common* property of consciousness. Defenders of divine timelessness have frequently pointed out that the act of knowing something need not take any time at all.[7] It makes sense, for example, to say that a timeless being knows the multiplication table. So why is an atemporal, conscious knowledge of unchanging truth impossible?

Gale responds that anyone who knows some particular truth must have a disposition to engage in certain temporal activities. But Gale's assertion is clearly false. There is no reason to think that God cannot know 2+2=4 without having a disposition to engage in temporal activities. And remember, on the Christian view, God is free to refrain from creation altogether, in which case I see no reason to think He must be disposed to engage in temporal activities at all.

I am not aware of any other arguments in the literature aimed at showing that an atemporal consciousness is impossible. Accordingly, we may conclude that no good reason has been given for thinking that God could not satisfy condition (ii) above. Similarly, condition (iii) is satisfied, since on the basis of our investigation thus far, I (and, I trust, the reader) can regard God, existing timelessly, as a being to whom a state of consciousness can be attributed. Again, even in our envisioned empty, timeless world, God is at least *capable* of regarding others as conscious—even if, were He to create such beings, He would not then be timeless. (We may leave that hypothesis an open question at this point.) Thus, God could satisfy condition (iv). What about condition (v)? God in the empty world is once more at least *capable* of verbal communication, for He could create language users like us and communicate to them by inspiring prophets or even causing sound waves in the thin air. Thus, (v) is met.

Could a timeless God be self-conscious, as (vi) stipulates? In order to be self-conscious a being must hold beliefs about himself not only from the third-person perspective, such as, in God's case, "God is omnipotent" or "God believes that 2+2=4," but also from the first-person perspective, such as "I

---

[6] Richard M. Gale, *On the Nature and Existence of God* (Cambridge: Cambridge University Press, 1991), 52.
[7] Nelson Pike, *God and Timelessness,* Studies in Ethics and the Philosophy of Religion (New York: Schocken, 1970), 124; Mann, "Simplicity and Immutability," 270; Paul Helm, *Eternal God* (Oxford: Clarendon, 1988), 64-65; John C. Yates, *The Timelessness of God* (Lanham, Md.: University Press of America, 1990), 173-174; Brian Leftow, *Time and Eternity,* Cornell Studies in the Philosophy of Religion (Ithaca, N.Y.: Cornell University Press, 1991), 285-290.

am omnipotent" or "I believe that 2+2=4."⁸ But it takes no more time to believe truly that "I have no human company," for example, than it does to believe that "No human beings exist." For any truth God knows from a third-person viewpoint, we can formulate a corresponding belief from the first-person perspective. Hence, if God can be timelessly conscious, there is no reason He cannot be timelessly self-conscious. Hence, criterion (vi) is also met.

That leaves criterion (i), that God must be rational in order to be personal. Without going into the debate over what it means to be rational, we may say rather confidently that God's being timeless impairs neither God's noetic structure (His system of beliefs) nor His ability to discharge any intellectual duties He might be thought to have. Since He is omniscient, it is pretty silly to think that God could be indicted for irrationality! Nor, as we have seen, would timelessness inhibit His knowing all truth in a timeless world such as we are contemplating.

Thus, a timeless God could fulfill all the various necessary conditions laid down for being personal. More than that, I should say that being self-conscious is not merely a *necessary* but also a *sufficient* condition for personhood. Our thought experiment of God's existing timelessly alone suggests that it is quite possible for God to be both timeless and self-conscious in such a state and, hence, personal.

Now some philosophers have denied that a timeless God can be a self-conscious, rational being, because He could not exhibit certain forms of consciousness which we normally associate with personal beings (namely, ourselves). The metaphysician Robert Coburn has written,

> Surely it is a necessary condition of anything's being a person that it should be capable (logically) of, among other things, doing at least some of the following: remembering, anticipating, reflecting, deliberating, deciding, intending, and acting intentionally. To see that this is so one need but ask oneself whether anything which necessarily lacked all of the capacities noted would, under any conceivable circumstances, count as a person. But now an eternal being would necessarily lack all of these capacities in as much as their exercise by a being clearly requires that the being exist in time. After all, reflection and deliberation take time, deciding typically occurs at some time—and in any case it always makes sense to ask, "When did you (he, they, etc.) decide?"; remembering is impossible unless the being doing the remembering has a past; and so on. Hence, no eternal being, it would seem, could be a person.⁹

---

⁸ Philosophers distinguish between knowledge *de re*, which is non-perspectival knowledge of a thing, and knowledge *de se*, which is self-knowledge.
⁹ Robert C. Coburn, "Professor Malcolm on God," *Australasian Journal of Philosophy* 41 (1963): 155.

Now even if Coburn were correct that a personal being must be capable of exhibiting the forms of consciousness he lists, it does not follow that a timeless God cannot be personal. For God could be *capable* of exhibiting such forms of consciousness but be timeless just in case (that is, "if and only if") He does not *in fact* exhibit any of them. In other words, the hidden assumption behind Coburn's reasoning is that God's being timeless or temporal is an essential property of God, that either God is necessarily timeless or He is necessarily temporal. But that assumption seems to me dubious. Suppose, for the sake of argument, that God is in fact temporal. Is it logically impossible that God could have been timeless instead? Since God's decision to create is free, we can conceive of possible worlds in which God alone exists. If He is unchanging in such a world, then on any relational view of time God would be timeless, as we have seen. In such an atemporal world God would lack certain properties which we have supposed Him to have in the actual world—for example, the property of *knowing what time it is* or the property of *co-existing with temporal creatures*—and He would have other properties which He lacks in the actual world—for example, the property of *being alone* or of *knowing that He is alone*—but none of these differences seems significant enough to deny that God could be either timeless or temporal and still be the same being. Just as my height is a contingent rather than essential property of mine, so God's temporal status is plausibly a contingent rather than essential property of His. So apart from highly controversial claims on behalf of divine simplicity or immutability, I see no reason to think that God is either *essentially* temporal or *essentially* timeless.

So if timelessness is a merely contingent property of God, He could be entirely capable of remembering, anticipating, reflecting, and so on; only were He to do so, then He would not be timeless. So long as He freely refrains from such activities He is timeless, even though He has the *capacity* to engage in those activities. Thus, by Coburn's own lights God must be regarded as personal.

At a more fundamental level, it is in any case pretty widely recognized that most of the forms of consciousness mentioned by Coburn are not essential to personhood—indeed, not even the capacity for them is essential to personhood. Take remembering, for example. Any temporal individual who lacked memory would be mentally ill or a mere animal. But if an individual exists timelessly, then he has no past to remember. He thus never forgets anything! Given God's omniscience, there is just no reason to think that His personhood requires memory. Similarly with regard to anticipation: Since a timeless God has no future, there just is nothing to anticipate. Only a temporal person needs to have beliefs about the past or future.

As for reflecting and deliberating, these are ruled out not so much by God's timelessness as by His omniscience. An omniscient being cannot reflect and deliberate because He already knows the conclusions to be arrived at! Even if God is in time, He does not engage in reflection and deliberation. But He is surely not impersonal as a result.

What about deciding, intending, and acting intentionally? I should say that all of these forms of consciousness are exhibited by a timeless God. With respect to deciding, again, omniscience alone precludes God's deciding in the sense of making up His mind after a period of indecision. Even a temporal God does not decide in that sense. But God does decide in the sense that His will intends toward one alternative rather than another and does so freely. It is up to God what He does; He could have willed otherwise. This is the strongest sense of libertarian freedom of the will. In God's case, because He is omniscient, His free decisions are either everlasting or timeless rather than preceded by a period of ignorance and indecision.

As for intending or acting intentionally, there is no reason to think that intentions are necessarily future-directed. One can direct one's intentions at one's present state. God, as the Good, can timelessly desire and will His own infinite goodness. Such a changeless intention can be as timeless as God's knowing His own essence. Moreover, in the empty world we have envisioned, God may timelessly will and intend to refrain from creating a universe. God's willing to refrain from creation should not be confused with the mere absence of the intention to create. A stone is characterized by the absence of any will to create but cannot be said to will to refrain from creating. In a world in which God freely refrains from creation, His abstaining from creating is a result of a free act of the will on His part. Hence, it seems to me that God can timelessly intend, will, and choose what He does.

Now some theologians have objected to the picture I have painted of a timeless, solitary deity, for such a being lacks all interpersonal relationships, and such relationships, they believe, are essential to personhood. If God is to be personal, He must be engaged in relationships with other persons. But the give-and-take of personal relationships inherently involves temporality.

In response to this objection, I think it would be extraordinarily difficult to prove that engaging in personal relationships, as opposed to the *capacity* to engage in personal relationships, is essential to personhood. A timeless God could have the capacity for such relationships even if, were He to engage in them, He would in that case be temporal. But let that pass. The more important assumption underlying this objection is the assumption that the persons to whom God is related must be *human* persons. For on the Christian conception of God, that assumption is false. Within the fullness of the Godhead itself, the persons of the Father, the Son, and the Holy Spirit enjoy the interpersonal relations afforded by the Trinity which God is. As a Trinity, God is eternally complete with no need of fellowship with finite persons. It is a marvel of God's grace and love that He would freely create finite persons and invite them to share in the love and joy of the inner Trinitarian life of God.

But would the existence of these Trinitarian interrelationships necessitate that God be temporal? I see no reason to think that the persons of the Trinity could not be affected, prompted, or responsive to one another in an unchanging and, hence, timeless way. To use a mundane example, think of iron filings clinging to a magnet. The magnet and the filings need not change their positions in any way in order for it to be the case that the filings are stuck to the magnet because the magnet is affecting them and they are responding to the magnet's force. Of course, on a deeper level change is going on constantly in this case because the magnet's causal influence is mediated by finite velocity electro-magnetic radiation. Nonetheless, the example is instructive because it illustrates how on a macroscopic level action and response can be simultaneous and, hence, involve neither change nor temporal separation. How much more is this so when we consider the love relationship between the members of the Trinity! Since intra-Trinitarian relations are not based on physical influences or rooted in any material substratum but are purely mental, the response of the Son to the Father's love implies neither change nor temporal separation. Just as we speak metaphorically of two lovers who sit, not speaking a word, gazing into each other's eyes as "lost in that timeless moment," so we may speak literally of the timeless mutual love of the Father, Son, and Spirit for one another.

The ancient doctrine of *perichoreisis*, championed by the Greek Church Fathers, expresses the timeless interaction of the persons of the Godhead.[10] According to that doctrine, there is a complete interpenetration of the persons of the Trinity, such that each is intimately bound up in the activities of the other. Thus, what the Father wills, the Son and Spirit also will; what the Son loves, the Father and Spirit also love, and so forth. Each person is completely transparent to the others. There is nothing new that the Son, for exam-

---

[10] See St. John Damascene, *An Exact Exposition of the Orthodox Faith* 2.1 (St. John of Damascus, *Writings* [New York: Fathers of the Church, 1958], 204).

ple, might communicate to the Spirit, since that has already been communicated. There exists a full and perfect exchange of the divine love and knowledge, so that nothing is left undone which needs to be completed. In this perfect interpenetration of divine love and life, no change need occur, so that God existing alone in the self-sufficiency of His being would, on a relational view of time, be timeless.

Thus, I think it is evident that God can enjoy interpersonal relations and yet be timeless. So even if we conceded that God is essentially timeless and that interpersonal relations are essential to personhood, it is still not true that if God is timeless, He cannot stand in interpersonal relations.

In conclusion, then, the argument for divine temporality based on God's personhood cannot be deemed a success. Advocates of a temporal God have not been able to show that God cannot possess timelessly the properties essential to personhood. On the contrary, we have seen that a timeless God can be plausibly said to fulfill the necessary and sufficient conditions of being a person. A timeless, divine person can be a self-conscious, rational individual endowed with freedom of the will and engaged in interpersonal relations.

All this has been said, however, in abstraction from the reality of a temporal universe. Given that such a universe exists, it remains to be seen whether God can remain untouched by its temporality.

II. Divine Relations with the World

Exposition

In the previous section we abstracted from the actual existence of the temporal world and considered God existing alone without creation and asked whether He could exist timelessly. We saw that He could. But, of course, the temporal world does exist. The question therefore arises whether God can stand in relation to a temporal world and yet remain timeless.

It is very difficult to see how He can. Imagine once more God existing changelessly alone without creation, but with a changeless determination of His will to create a temporal world with a beginning. Since God is omnipotent, His will is done, and a temporal world comes into existence. Now this presents us with a dilemma: Either God existed prior to creation or He did not. Suppose He did. In that case, God is temporal, not timeless, since to exist *prior* to some event is to be in time. Suppose, then, that God did not exist prior to creation. In that case, without creation, He exists timelessly, since He obviously did not come into being along with the world at the moment of creation.

This second alternative presents us with a new dilemma: Once time begins at the moment of creation, either God becomes temporal in virtue of His real relation to the temporal world or else He exists just as timelessly with creation as He does without it. If we choose the first alternative, then, once again, God is temporal. But what about the second alternative? Can God remain untouched by the world's temporality? It seems not. For at the first moment of time, God stands in a new relation in which He did not stand before (since there was no "before"). Even if in creating the world God undergoes no *intrinsic* change, He at least undergoes an *extrinsic* change.[11] For at the moment of creation, God comes into the relation of *sustaining* the universe or, at the very least, of *co-existing with* the universe, relations in which He did not stand before. Since He is free to refrain from creation, God could have never stood in those relations, had He so willed. But in virtue of His creating a temporal world, God comes into a relation with that world the moment it springs into being. Thus, even if it is not the case that God is temporal prior to His creation of the world, He nonetheless undergoes an extrinsic change at the moment of creation which draws Him into time in virtue of His real relation to the world. So even if God is timeless without creation, His free decision to create a temporal world also constitutes a free decision on His part to exist temporally.

The argument of the advocate of divine temporality can be summarized as follows:

1. God is creatively active in the temporal world.

2. If God is creatively active in the temporal world, God is really related to the temporal world.

3. If God is really related to the temporal world, God is temporal.

4. Therefore, God is temporal.

This argument, if successful, does not prove that God is essentially temporal, but that if He is a Creator of a temporal world—as He in fact is—then He is temporal.

---

[11] Recall the distinction made in chapter 2, note 2, between intrinsic and extrinsic change. It is disputed among philosophers of religion whether creating the world involves some intrinsic change on God's part (for example, an exercise of power). My argument does not presuppose an intrinsic change in God but is based on the inevitability of mere extrinsic change on God's part.

CRITIQUE

One way to escape this argument is to deny premise (2). This might not appear to be a very promising strategy, since it seems obvious that God is related to His creatures insofar as He sustains them, knows them, and loves them. Remarkably, however, it was precisely this premise that medieval theologians such as Thomas Aquinas denied.

Thomas agrees with premise (3). On his view, relations between God and creatures, such as God's being Lord over the world, first begin to exist at the moment at which the creatures come into being. Hence, if God stands in real relations to His creatures, He acquires those relations new at the moment of creation and thus undergoes extrinsic change. And anything that changes, even extrinsically, must be in time.

Thomas escapes the conclusion that God is therefore temporal by denying that God stands in any real relation to the world. Since God is absolutely simple, He stands in no relations to anything, for relations would introduce complexity into God's being. Aquinas holds, paradoxically, that while creatures are really related to God, God is not really related to creatures. The relation of God to creatures is just in our minds, not in reality.

To give an illustration: Suppose Joe is jealous of John. In that case, Joe is related to John by the *is envious of* relation, and John is related to Joe by the *is envied by* relation. But Aquinas would say that only Joe's relation to John is real: He really is envious of John. But John's relation to Joe is just in our heads: Whether Joe exists or not, John is the same; his being envied by Joe does not make any real difference in him.

Similarly, Aquinas says, creatures are really sustained, known, and loved by God, but God would be the same whether creatures existed or not. He therefore does not stand in real relations of sustaining, knowing, or loving His creatures. On Aquinas's view, then, God undergoes no extrinsic change in creating the world. He just exists, and creation is creatures' coming into existence with a real relation to God of being caused by God.

This is certainly an extraordinary doctrine. Wholly apart from its reliance on divine simplicity, the doctrine of no real relations is very problematic. God's sustaining the world is a causal relation rooted in the active power and intrinsic properties of God as First Cause. It is therefore not at all analogous to the passive relation *is envied by*. Thus, to say that the world is really related to God by the relation *is sustained by,* but that God is not really related to the world by the relation *is sustaining* seems unintelligible. It is to say that one can have real effects without a real cause—which seems self-contradictory or incomprehensible.

Moreover, God is surely really related to His creatures in the following sense: In different logically possible worlds which we can imagine, God's will, knowledge, and love would be different than they actually are. For example, if God had not chosen to create a universe at all, He would surely have a different will than that which He has (for He would not will to create the universe); He would know different truths than the ones He knows (for example, He would not know *The universe exists,* since that would be false in that world); He would not love the same creatures He actually loves (since no creatures would exist). Incredibly, however, Aquinas denies this. It is the implication of his view that God is perfectly similar in every possible world we can conceive: He never wills differently, He never acts differently, He never knows differently, He never loves differently. Whether the world is empty or chock-full of creatures of every sort, there is no difference in God. But then it becomes unintelligible why this universe or any universe exists rather than just nothing. The reason cannot lie in God, for He is perfectly similar in all possible worlds. Nor can the reason lie in creatures, for we are asking for some explanation of their existence. Thus, on Thomas's view there just is no reason for why this universe or any universe at all exists.

Therefore, Thomas's attempt to evade the present argument by denying premise (2) is just not plausible. The defender of divine timelessness must seek some other way of escape.

Recent defenders of timeless eternity have turned their guns on premise (3) instead. They have tried to craft theories of divine eternity that would permit God to be really related to the temporal world and yet to exist timelessly.

For example, Eleonore Stump and the late Norman Kretzmann, who rekindled contemporary discussion of divine timelessness, attempted to craft a new simultaneity relation, which they believed would allow a timeless God to relate to His creation.[12] They understand the generic relation of simultaneity to be *existence at once* (or *together*). Temporal simultaneity is one type of simultaneity indicating *existence at one and the same time.* Eternal simultaneity (which would hold between timeless entities, say, God and numbers) is *existence at one and the same eternal present.* Now the problem of relating a timeless entity to a temporal entity is that there is no single mode of existence that would allow one to define Eternal-Temporal simultaneity as

---

[12] Eleonore Stump and Norman Kretzmann, "Eternity," *Journal of Philosophy* 78 (1981): 429-458.

existence at one and the same _____. There is nothing to fill in the blank. So how can one relate two such disparate modes of existence as timelessness and temporality?

In order to craft a definition of this new type of simultaneity (which they abbreviate as ET-simultaneity), Stump and Kretzmann appeal to the analogy of the Special Theory of Relativity (STR). There, as we have seen, simultaneity is relative to inertial frames. Temporal simultaneity means *existence at one and the same time within the reference frame of a given observer.* Stump and Kretzmann propose to treat modes of existence as analogous to reference frames and to construct a definition of ET-simultaneity in terms of two reference frames (timelessness and temporality) and two observers (one in eternity and one in time).

Their definition is very complicated in its wording, but the basic idea is as follows. Take some eternal being x and some temporal being y. These two are ET-simultaneous just in case, relative to some hypothetical observer in the eternal reference frame, x is eternally present and y is observed as temporally present, and relative to some hypothetical observer in any temporal reference frame, y is temporally present and x is observed as eternally present.

A word of clarification: By "eternal" Stump and Kretzmann mean "timeless," and by "temporal reference frame" they mean "moment of time." It is also worth noting that this definition is not really analogous to simultaneity in STR at all. A better analogy would be to say that x and y are ET-simultaneous just in case they both exist at the same eternal present relative to the eternal reference frame and both exist at the same moment of time relative to the temporal reference frame. But then God would be temporal relative to our mode of existence, which Stump and Kretzmann do not want to say.

On the basis of their definition of ET-simultaneity, Stump and Kretzmann believe they have solved the problem of how a timeless being can be really related to a temporal world. For relative to the eternal reference frame, any temporal entity which exists at any time is observed to be present, and relative to any moment of time God is observed to be present. The metaphysical relativity postulated by ET-simultaneity implies that all events are present to God in eternity and therefore open to His timeless causal influence. Every action of God is ET-simultaneous with its temporal effect.

Now the Stump-Kretzmann account is a veritable mare's nest of philosophical difficulties. But in the interest of brevity, let us pass them by and cut to the heart of the matter: Their proffered definition of ET-simultaneity is explanatorily vacuous. As many critics have pointed out, the language of observation employed in the definition is wholly obscure.[13] In STR very specific physical content is given to the notion of observation through Einstein's operational definitions of distant simultaneity. But in the definition of ET-simultaneity, no hint is given as to what is meant, for example, by x's being observed as eternally present relative to some moment of time. In the absence of any procedure for determining ET-simultaneity, the definition reduces to the assertion that, relative to the reference frame of eternity, x is eternally present and y is temporally present, and that relative to some temporal reference frame, y is temporally present and x is eternally present—which is only a restatement of the problem! Worse, if y is temporally present to God, then God and y are not ET-simultaneous at all, but temporally simultaneous. Thus, God would be temporally simultaneous with every temporal event, which is to sacrifice divine timelessness.

Paul Helm of the University of London, himself a defender of divine timelessness, is not being uncharitable when he complains that Stump and Kretzmann's "'solution' to the problem is found simply by rewording the problem with the help of the device of ET-simultaneity. ET-simultaneity has no independent merit or use, nothing is illuminated or explained by it."[14]

To their credit, Stump and Kretzmann later revised their definition of ET-simultaneity so as to free it from observation language.[15] Basically, their new account tries to define ET-simultaneity in terms of causal relations. On the new definition, x and y are ET-simultaneous just in case, relative to an observer in the eternal reference frame, x is eternally present and y is temporally present, and the observer can enter into direct causal relations with both x and y; and relative to an observer in any temporal reference frame, x is eternally present and y is at the same time as the observer, and the observer can enter into direct causal relations with both x and y.

Again, there are many difficulties with this new definition which we may overlook. The fundamental problem with this new account of ET-simultaneity is that it is viciously circular. For ET-simultaneity was originally invoked to explain how a timeless God could be causally active in time; but now ET-simultaneity is defined in terms of a timeless being's ability to be causally active in time. Our original problem was to explain how God could be both

---

[13] Stephen T. Davis, *Logic and the Nature of God* (Grand Rapids, Mich.: Eerdmans, 1983), 20; Delmas Lewis, "Eternity Again: A Reply to Stump and Kretzmann," *International Journal for Philosophy of Religion* 15 (1984): 74-76; Helm, *Eternal God*, 32-33; William Hasker, *God, Time, and Knowledge* (Ithaca, N.Y.: Cornell University Press, 1989), 164-166; Yates, *Timelessness of God*, 128-30; Leftow, *Time and Eternity*, 170-172.

[14] Helm, *Eternal God*, 33.

[15] Eleonore Stump and Norman Kretzmann, "Eternity, Awareness, and Action," *Faith and Philosophy* 9 (1992): 477-478.

timeless and yet creatively active in the world. That is hardly explained by saying that a timeless God is ET-simultaneous with His effects in time and then defining ET-simultaneity in terms of the ability of a timeless being to be causally related to temporal effects. This amounts to saying that God can be causally active in time because He can be causally active in time! Brian Leftow, who has written extensively on God and time, concludes,

> any definition of ET-simultaneity which invokes any form of ET-causality ... is implicitly circular. For to fully explain how ET-causation can occur, we must bring in the concept of ET-simultaneity. If we do, we cannot then define ET-simultaneity by invoking ET-causation, for then the concept to be defined in effect recurs in the definition.[16]

Since their first definition was explanatorily vacuous and their second definition viciously circular, Stump and Kretzmann must be judged to have failed in their attempt to undercut premise (3) of the argument under discussion and so to stave off its conclusion.

Leftow himself has offered another, different account of divine eternity in order to refute premise (3).[17] It will be recalled that on the Stump-Kretzmann model, there is no common reference frame or mode of existence shared by timeless and temporal beings. As a result, Stump and Kretzmann were unable to explain how such beings could be causally related. The essence of Leftow's proposal is to remedy this defect by maintaining that temporal beings do exist in eternity; they share God's mode of existence and so can be causally related to God. But, he insists, this does not imply that time or temporal existence is illusory, for temporal beings also have a temporal mode of existence.

How can it be shown that temporal beings exist in timeless eternity? Leftow's argument is based on three theses:

I. The distance between God and every thing in space is zero.

II. Spatial things do not change in any way unless there is a change of place (a motion involving a material thing).

III. If something is in time, it is also in space.

---

[16] Leftow, *Time and Eternity*, 173.
[17] Brian Leftow, "Eternity and Simultaneity," *Faith and Philosophy* 8 (1991): 148-179; cf. idem, *Time and Eternity*, chapter 10.

On the basis of these theses Leftow argues as follows: There can be no change of place relative to God because the distance between God and everything in space is zero. But if there is no change of place relative to God, there can be no change of any sort on the part of spatial things relative to God. Moreover, since anything that is temporal is also spatial, it follows that there are no temporal, non-spatial beings. The only temporal beings there are exist in space, and none of these changes relative to God. Assuming, then, some relational view of time, according to which time cannot exist without change, it follows that all temporal beings exist timelessly relative to God. Thus, relative to God all things are timelessly present and so can be causally related to God.

The problem with this reasoning is that all three of its foundational theses seem false, some obviously so. Take (I), for example. This thesis rests pretty obviously on a category mistake. When we say that there is no distance between God and creatures, we do not mean that there is a distance and its measure is zero. Rather we mean that the category of distance does not even apply to the relations between a non-spatial being such as God and things in space.

There is a helpful illustration of the point from the history of Relativity Theory. In defending the existence of an aether, H. A. Lorentz had denuded the aether of virtually every physical property except the property of being motionless; the aether as he conceived it was virtually equivalent to the inertial frame of absolute space. Einstein once joked that all he had done was to divest the aether of the last physical property Lorentz had left it: its state of motion.[18] Einstein said that he was quite willing to admit the existence of an aether just so long as no state of motion is ascribed to it. Now in denying that the aether has a state of motion, Einstein was clearly not saying that the motion of the aether was zero. That was Lorentz's position: The aether is at rest; it has a state of motion and its measure is zero. What Einstein was saying is that the category of motion does not even apply to the aether: It is neither moving nor at rest. To think that it is is a category mistake.

Now in exactly the same way, thesis (I) is just a category mistake. The concept of spatial separation or distance cannot be applied to a being which transcends space and things in space. It is therefore wrong-headed to say that the distance between them is zero. Unfortunately, Leftow's whole theory balances on thesis (I) like a pyramid on its point. Without it, the theory collapses because things in space are not, then, changeless relative to God.

But let us press on. What about thesis (II)? Again, this thesis is false if

---

[18] A. Einstein, *Äther und Relativitätstheorie* (Berlin: Julius Springer Verlag, 1920), 7-9.

time is dynamic or "tensed."[19] On this view of time, the difference between past, present, and future is not just in our minds, and temporal becoming is real. If temporal tenses are real, spatial things can change even if there is no spatial motion by changing in their temporal properties. For example, some spatial object can change by being one year old and then becoming two years old, even if no change of place has occurred. (It will be recalled that even most relationalists are today willing to admit that time can go on during periods of spatial changelessness.[20]) The significance of the falsity of (II) is that, even if the entire universe were frozen into immobility, there would still be change relative to God, namely, change of temporal properties. Hence things would not be changeless and therefore timeless relative to God, which undermines Leftow's claim that temporal beings exist in eternity. Thus, if time is dynamic—and Leftow allows that it may be—then his theory is nullified.

Finally, consider thesis (III). Leftow needs this thesis, lest someone say that there are non-spatial, temporal beings such as angels that are changing relative to God. Such beings would (on Leftow's analysis) have a zero distance from God and yet not be changeless relative to God. Thus, they would not exist in eternity. So in order to sustain his claim that temporal beings exist in eternity, Leftow has to get rid of such beings. He does so by means of the reductionistic thesis (III), which says that if anything exists in time it also exists in space.

Now we have already seen good reason to reject this radical thesis.[21] Even in the absence of a physical universe, God could choose to entertain a succession of thoughts or to create an angelic being or an unembodied soul which experiences a stream of consciousness, and such a series of mental events alone is sufficient for such entities' being in time. On what grounds, then, does Leftow adopt (III)?

Leftow appeals to the geometric representation of space-time in contemporary physics as a justification for (III).[22] In such a geometrical presentation, three dimensions of the geometry represent length, width, and height, and the fourth represents time. If something has a coordinate in one dimension of this four-dimensional structure, then it has three other coordinates as well. Thus, if something is in time, it must be in space as well.

There is a huge metaphysical assumption underlying this reasoning, however, a veritable philosophical iceberg of which Leftow seems unaware: the

---

[19] Such a view is in contrast to the static or tenseless view of time mentioned in chapter 2, pages 69-70.
[20] See note 2 above.
[21] See chapter 2, page 66.
[22] For more on the geometrical representation of space-time, see chapter 5, pages 167-180.

assumption of space-time realism. That is to say, his reasoning presupposes that the geometrical representation of space-time is more than just a graphic way of presenting STR or GTR—that it depicts the actual structure of the world. It is to suppose that temporal becoming is unreal; that things located at any spatio-temporal location are equally real or existent. In other words, space-time realism entails the static or "tenseless" view of time alluded to earlier.[23] We shall have much to say of this later; but for now it is sufficient to note that such a metaphysical assumption requires some justification.

Neither STR nor GTR requires space-time realism, for in the original 1905 STR paper Einstein treated time as a parameter, not a coordinate; that is to say, he did not assume a four-dimensional view of the world, a view he came to adopt only later under the influence of the mathematician Hermann Minkowski. Similarly, in GTR-based cosmological models, cosmic time is a parameter, not a coordinate.

A good many philosophers of science think of the four-dimensional, geometrical representation of space-time, not realistically, but *instrumentally,* that is to say, as an elegant and handy way of presenting STR or GTR and of thinking about problems of time and space; but they do not invest such pictures with reality. For example, the French physicist Henri Arzeliès writes, "The Minkowski continuum is an abstract space of four-dimensions, the sole role of which is to interpret in geometrical language statements made in algebraic or tensor form. . . . The four-dimensional continuum should therefore be regarded as a useful tool, and not as a physical 'reality'."[24] Similarly, philosopher Max Black complains, "this picture of a 'block universe,' composed of a timeless web of 'world lines,' in a four-dimensional space, however strongly suggested by the theory of relativity, is a piece of gratuitous metaphysics."[25] Some reason is needed, therefore, if we are to reject an instrumentalist view of space-time in favor of a realist interpretation. And that Leftow has not given.

Now it might be said that even on a dynamic view of time according to which only the present exists, still if things are in time, they must be in space. But even leaving aside the distinction between parameter time and coordinate time, what we must keep in mind is that such an assertion assumes that Newton erred in distinguishing between time itself and our measures of time.

---

[23] See chapter 2, pages 69-70.
[24] Henri Arzeliès, *Relativistic Kinematics*, rev. ed. (Oxford: Pergamon Press, 1966), 258. The mathematics of STR is algebra; the mathematics of GTR is called tensor calculus.
[25] Max Black, review of *The Natural Philosophy of Time*, by G. J. Whitrow, *Scientific American* 206 (April 1962), 181.

Even if our *measures* of time and space are bound up together, that is no reason to think that time and space themselves cannot exist independently. On the contrary, we have seen good reason to think that they can, since mental events alone are a sufficient condition of a temporal series. Thus, things can exist temporally without existing spatially.

In short, Leftow's thesis (III) assumes both space-time realism and the identity of time and space with our physical measures thereof—enormous assumptions which we have good reason to doubt.

Thus, all of Leftow's key theses are at least dubious, if not clearly false. We have little choice but to conclude that he has given no good grounds for thinking that temporal beings exist in timeless eternity.

Moreover, we must ask, is Leftow's theory even coherent? If all events exist timelessly in God's eternal reference frame, then none of them can exist *earlier than, simultaneous with,* or *later than* another event, for these are temporal relations. Thus, in God's reference frame, all He is confronted with is a chaos of point-events all temporally unrelated to one another. This not only seems incompatible with divine omniscience and providence but contradicts Leftow's own statements that in eternity God discerns the sequence in which events occur.

Finally, if all things really do exist timelessly in eternity, are not time and temporal existence ultimately illusory? Leftow denies this because in STR actuality, like simultaneity, is relative to inertial frames. Thus, things can be actual relative to God's reference frame but not yet actual or no longer actual relative to the temporal reference frame.

It is worth noting two things about this appeal to STR. First, while one *can* relativize actuality to inertial frames or space-time points in STR, doing so has enormously implausible consequences (of which we shall speak later[26]). These might well prompt us to prefer a Lorentzian interpretation of STR instead. In that case, all frames are not relative, as Leftow's theory demands. Second, in STR simultaneity relations are *not* relative for causally connected events. For causally connected events the relations *earlier than, simultaneous with,* and *later than* are absolute. Since God is causally related to all events, His timeless relation with them should be absolute, not relative. Insofar as Leftow must deny this, his theory is disanalogous to STR.

In any case, serious objections may be lodged against Leftow's metaphysical relativity. (1) God's frame of reference is surely privileged. As the Creator of the universe, God's timeless mode of existence in which He sus-

---

[26] See chapter 5, pages 169-173.

tains all events ought to be recognized as the preferred frame. In that case, Leftow's theory implies that time and temporal becoming are illusions of finite creatures, who are ultimately timeless in their being. (2) If we deny the preferred status of God's frame and insist on a democracy of frames, then relative to the temporal reference frame God ought to be in time, just as relative to the eternal frame creatures are timeless. For in the temporal frame of reference, God undergoes extrinsic change in virtue of the intrinsic changes in creatures to whom He is related. (3) In any case, the metaphor of God's frame of reference is empty, being based on the spurious assumption (I) above. God, as a non-spatial object, simply is not spatially related to creatures and so has no "reference frame" as such. If one means by this metaphor simply His timeless mode of existence, then it does not seem logically coherent to speak of temporal beings' sharing a timeless mode of existence. How can creatures be coherently said to exist both timelessly and temporally? It explains nothing to appeal to metaphors of reference frames relative to which creatures are timeless or temporal, for these reference frames just are the two modes of existence, timelessness and temporality, and it merely restates the problem to say that creatures exist both ways. Thus, Leftow's theory proves no more successful than Stump and Kretzmann's in explaining how God can be timeless and yet causally related to the world.

In summary, it seems to me that we have here a powerful argument for divine temporality. Classical attempts such as Aquinas's to deny that God is really related to the world, and contemporary attempts such as those of Stump, Kretzmann, and Leftow to deny that God's real relation to the world involves Him in time, all appear in the end to be less plausible than the premises of the argument itself. It seems that in being related to the world God must undergo extrinsic change and so be temporal.

III. Divine Knowledge of Tensed Facts

Exposition

We have seen that God's real relation to the temporal world gives us good grounds for concluding God to be temporal in view of the extrinsic change He undergoes through His changing relations with the world. But the existence of a temporal world also seems to entail intrinsic change in God in view of His knowledge of what is happening in the temporal world. For since what is happening in the world is in constant flux, so also must God's knowledge of what is happening be in constant flux. Defenders of divine temporality have argued that a timeless God cannot know certain tensed facts about the world—for example, what is happening now—and therefore, since God is omniscient, He must be temporal.

With this argument we move out of the philosophy of science and into the philosophy of language. The key notion to be understood here is the idea of "tensed facts."

First, let us say a word about what we mean by "fact." A fact may be defined as the state of affairs described by a true declarative sentence.[27] Thus, for example, while "Snow is white" and "Der Schnee ist weiß" are two different sentences, they both describe the same fact, namely, snow's being white.

Second, let us define what we mean by a "tensed fact." We are all familiar with tense as it plays a role in language. In English we normally express tense by inflecting the verb of a sentence so as to express the past, present, or future tense, or by compounding verbs to express more complex tenses such as the past perfect or the future perfect. Although most of our ordinary language is tensed, there are occasions on which we employ sentences that are grammatically in the present tense to express what are really tenseless truths. For example, we say such things as "Lady Macbeth commits suicide in Act V. scene v," "The glass breaks easily," "The area of a circle is πr^2," and "Centaurs have the body of a horse and the torso of a man." That the verbs in the above sentences are in fact tenseless is evident from the fact that it would be wrong-headed to replace them by the present tense equivalent of "is + (present participle)," for example, "is committing," "is breaking," and so forth. Such a substitution would render some of these true sentences plainly false.

The function of tense is to locate something in relation to the present. This can be done not only by means of verbs but also by means of temporal indexical expressions. An indexical expression is a word or phrase which systematically changes its referent (the thing it refers to) as the context of its utterance changes. For example, if I tell someone on June 1, "John is arriving tomorrow," the indexical word "tomorrow" refers to June 2. But on the next day, if I were to say, "John is arriving tomorrow," the same word would refer, not to June 2, but to June 3. In order to refer to June 2, I should have to employ a different indexical word, "today," to express John's arriving on that day. The reason these expressions systematically change their referents is due to their being tensed expressions. They locate something in relation to the present, which is constantly changing, and so what they refer to changes, too.

---

[27] What I am calling a fact could be treated as a true proposition. Accordingly, what I call "factual content" is the same as "propositional content." I am trying my best to avoid technical jargon.

Temporal indexical expressions include adverbial phrases (such as "today," "now," "three days ago"), adjectives (such as "past," "present," and "future"), prepositional phrases (such as "by next Saturday," "at present," "in yet two days' time"), and even nouns (as in "Today is Wednesday").

Such tensed expressions differ radically from expressions using clock-times or dates, which are tenseless. "January 3, 1812" invariably refers to the same day, whether it is past, present, or future; whereas temporal indexical expressions such as "yesterday," "today," or "tomorrow" depend upon the context of their utterance for what day they refer to. Dates can therefore be employed in conjunction with tenseless verbs to locate things tenselessly in time. For example, we can state, "In 1960 John Kennedy *pledges* to send a man to the moon before the end of the decade" (the italics being a stylistic convention to show that the verb is tenseless). This sentence expresses a tenseless fact and is therefore always true. Notice that even if one knew this truth, one would not know whether Kennedy has issued his pledge unless one also knew whether 1960 was past or future. By contrast, if we replaced the tenseless verb with the past-tensed verb "pledged," then we would know that the event referred to has happened. This tensed sentence would, however, not always be true: Prior to 1960 it would be false. Prior to 1960 the tensed verb would have to be the future-tense "will pledge" if the sentence is to be true. In contrast to tenseless sentences, then, tensed sentences serve to locate things in time relative to the present and so may change their truth value.

The salient point of all this is that in addition to tenseless facts, there also appear to be tensed facts. The information conveyed by a tensed sentence concerns not just tenseless facts but also tensed facts as well, facts about how far from the present something is. Thus, what is a fact at one moment may not be a fact at another moment. It is now a fact that I am writing this sentence; in a moment it will no longer be a fact. Thus the body of tensed facts is constantly changing.

The upshot is that a being which only knew all tenseless facts about the world, including which events *occur* at any date and time, would still be completely in the dark about tensed facts. He would have no idea at all of what is now going on in the universe, of which events are past and which are future. On the other hand, any being which *does* know tensed facts cannot be timeless, for his knowledge must be in constant flux, as the tensed facts known by him change.

Thus we can formulate the following argument for divine temporality:

1. A temporal world exists.

2. God is omniscient.

3. If a temporal world exists, then if God is omniscient, God knows tensed facts.

4. If God is timeless, He does not know tensed facts.

5. Therefore, God is not timeless.

Again, this argument does not prove that God is essentially timeless, but, if successful, it does show that if a temporal world exists, then God is not timeless.

Critique

Defenders of divine timelessness have attempted to refute this argument either by arguing that a timeless God can know tensed facts or by revising the definition of omniscience, so that God may still qualify as omniscient even if He is ignorant of tensed facts.

Let us look first at the plausibility of denying premise (4). Can a timeless God know tensed facts? Jonathan Kvanvig, a keen philosophical thinker at Texas A & M, contends that He can.[28] Kvanvig's defense of this point is based upon his analysis of beliefs in terms of a personal attitude, the factual content of a belief, and a particular way of accessing or grasping that factual content. Take a sentence such as "Today is June 1, 1999." Kvanvig contends that the same factual content is expressed by the sentence "Today is June 1, 1999," when that sentence is used on June 1, 1999, as is expressed by the sentence "Yesterday was June 1, 1999," when that sentence is uttered on June 2, 1999. In his view, temporal indexical words express the individual essence of the moment they refer to (an essence being a set of properties which uniquely designate a thing). In this example, the words "today" and "yesterday," by expressing the essence of the moment referred to, pick out the same time. A person will grasp this factual content *directly* only if he grasps it at the very time referred to (in which case he will form a present-tense belief), and a person will grasp the same content *indirectly* if he does so not at that time (in which case a temporal person will form beliefs involving other tenses).

---

[28] Jonathan L. Kvanvig, *The Possibility of an All-Knowing God* (New York: St. Martin's, 1986), 150-165.

In God's case, then, if He is timeless, He grasps the factual content of tensed sentences indirectly and so does not form tensed beliefs as we do, who grasp some factual content directly. Therefore, Kvanvig concludes, "one can affirm the doctrines of timelessness, immutability and omniscience by affirming that God indirectly grasps every temporal moment, and directly grasps none of them."[29]

Does Kvanvig's theory succeed in giving an account of how a timeless God can know tensed facts? It seems not. For on Kvanvig's analysis the essences of the times picked out by temporal indexical words do not include the tense of those times (that is, whether they are past, present, or future). Otherwise a time would be, say, essentially past, in which case it is impossible for that time ever to have been present or future, which is absurd. Words such as "today" and "yesterday" could not refer to the same day, since they, being different in tense, would express different essences. And God could not timelessly grasp the factual content involving such essences, since if He grasped a moment which is essentially present, He would exist at the time of that moment. Thus, it is evident that the factual content expressed by tensed sentences is, on Kvanvig's analysis, tenseless. Tense is merely a feature of our mental state, the by-product of how we grasp the tenseless factual content of tensed sentences. Kvanvig explicitly denies that there is any temporal element expressed by tensed sentences which is not part of their factual content. Thus, on Kvanvig's view tense is merely linguistic: There are no tensed facts.

Thus, Kvanvig's account backfires. Far from explaining how a timeless God can know tensed facts, on his analysis there are no tensed facts to be known. The factual content expressed by the sentence "Kvanvig now teaches at Texas A & M" is something like *Kvanvig teaches* (tenselessly) *at Texas A & M at time t*. God, grasping this factual content indirectly, has no idea where Kvanvig is now teaching or whether he has even been born or is long dead and buried.

A somewhat similar, but crucially modified, account of God's knowledge of tensed facts has been offered by Edward Wierenga in his philosophical analysis of the principal divine attributes.[30] On Wierenga's view the factual content of a present-tense sentence includes the tense expressed in the sentence. Like Kvanvig, he believes that moments of time have individual essences. Unlike Kvanvig, however, Wierenga seems to believe that the individual essence of a moment somehow involves the present tense. If the fac-

---

[29] Ibid., 159.
[30] Edward R. Wierenga, *The Nature of God: An Inquiry into Divine Attributes*, Cornell Studies in Philosophy of Religion (Ithaca, N.Y.: Cornell University Press, 1989), 179-185.

tual content of a sentence includes a moment's individual essence, then that content will involve presentness. Anyone who grasps that content at the time referred to will form a present-tense belief about what is "now" the case.

Wierenga contends that a timeless God is able to grasp the factual content of a tensed sentence, but without forming a present-tense belief as we do. For one only forms a present-tense belief if one both exists at the time referred to in the factual content of a sentence and grasps that content at that time. God grasps this factual content timelessly and so forms no present-tense belief about what is "now" going on. Thus, God knows tensed facts without having tensed beliefs.

Does Wierenga's account of God's knowledge of tensed facts fare any better than Kvanvig's analysis? The crucial difference between them is that Wierenga makes presentness a feature of the individual essence of every moment of time. At first blush this might seem hopelessly incoherent. If the essence of *every* moment of time involves presentness, then every moment of time would be present and no moment would ever be future or past, which makes nonsense of time. Wierenga could escape this absurdity, however, by advocating a view of time called *presentism,* according to which the only time which exists is the present time. According to presentism, future times do not yet exist and past times no longer exist. Therefore, there literally are no times which have the properties of pastness or futurity. When a time becomes past, it does not exchange the property of presentness for the property of pastness; rather it just ceases to exist altogether. Thus, the individual essences of all moments of time could involve presentness. This does not imply that all moments of time are somehow permanently present. Rather it implies that times are present when and only when they exist. They come into existence successively and are present just as long as they exist. No time exists which is not present, but that does not imply that all times are present together.

If the individual essence of every moment of time somehow involves presentness, then the question is whether a timeless God can grasp the factual content involving such an essence. It remains extraordinarily difficult to understand on Wierenga's account how God can grasp the essence of a time without that time's being present for Him. Consider the analogy (which Wierenga himself appeals to) of first-person indexical words such as "I". If my individual essence is, as Wierenga says, the property of just *being me,* then how can God possibly grasp the factual content of a sentence which includes this essence? If I say, "I feel miserable," then God could grasp the fact that *William Craig feels miserable;* but if the factual content of this sentence includes my individual essence of *being me,* expressed by the word "I," then a "Private Access Only" sign is posted before the route to this factual content, which is open to me alone. Even God could not grasp this content, since He and I are not the same person.

Analogously, if the individual essence of a time involves presentness, then in order to grasp the factual content of a sentence involving such an essence one would need to be present. If I say, "John left three hours ago," then there is no problem in a timeless God's grasping factual content involving the time t and the property *being such that John leaves three hours earlier than then,* which is ascribed to t—no problem, that is, so long as t is a tenseless date or clock-time. But if t involves presentness, then God, in grasping t as present, must be in the present, that is to say, must be temporal. Later at t' it will be true that "John left four hours ago," and God will no longer grasp the essence of t, but of t', for t is no longer present. It is always true that John *leaves* three hours earlier than t, and God immutably knows that fact. But if He is to know tensed facts, He must know that t is present. Thus, His factual knowledge must be constantly changing, in which case God must be in time. Hence, in making presentness part of the individual essence of every time, Wierenga only succeeds in temporalizing God.

Finally, let us consider Brian Leftow's account of God's timeless knowledge of tensed facts.[31] It will be recalled that on Leftow's view all events exist in eternity, where they are eternally actual, even though in time these same events are past, present, or future. Thus, relative to God's frame of reference, it seems that there are no tensed facts to be known. Relative to eternity no event is known as temporally past, present, or future, for all events are timelessly present to God.

But Leftow maintains that God also knows events as they exist relative to various temporal frames of reference. Leftow says that God knows "all the facts of simultaneity" relative to any temporal reference frame.[32] Thus, God would know that relative to the temporal frame R_1, events e_1 and e_2 are simultaneous, whereas relative to R_2, e_1 is earlier than e_2. But clearly such knowledge does not constitute knowledge of tensed facts, as Leftow believes. It is merely knowledge of the classes of simultaneous events at any time relative to any reference frame. Since simultaneity is a purely tenseless relation, it will not yield knowledge of what is past, present, or future relative to any frame. God could at best know relative to any frame the tenseless location of any event (its date and time) and its tenseless relation to any other event

---

[31] Leftow, *Time and Eternity,* 312-337.
[32] Ibid., 334.

(whether it is *earlier than, simultaneous with,* or *later than*). But He would not know, even relative to a temporal frame of reference, what events are past, present, or future. (If He did, His knowledge concerning that frame would be constantly changing, and thus He would exist in the time of that frame.)

Indeed, Leftow's explanation of what He calls "factual omniscience" makes it evident that on his view God's knowledge is tenseless. For he contends that the same fact renders true *It is now 3:00* and *It is then 3:00* at their respective times. The fact which makes these statements true can be grasped independently of the times at which they are true. Thus, he concludes, God can be factually omniscient even if He cannot grasp tensed truths.

This explanation reveals that on Leftow's view there really are no tensed facts. For a tensed fact such as *its now being 3:00* can only be grasped at 3:00. One can timelessly grasp *its being 3:00 at 3:00* or *its being 3:00 earlier than 4:00*, but such knowledge leaves one completely in the dark as to whether 3:00 is past, present, or future. Since on Leftow's account God knows all facts, and the only temporal facts known to God are tenseless facts, it follows that there are no tensed facts. Thus, Leftow's account of God's knowledge fails to supply Him knowledge of tensed facts.

Kvanvig, Wierenga, and Leftow's accounts are the most sophisticated attempts to explain how God can be timeless and yet know tensed facts, yet they all fail. Thus, premise (4) of the argument for divine temporality from God's knowledge of tensed facts seems secure.

The defender of divine timelessness has no recourse, then, but to deny premise (3). He must deny that omniscience entails a knowledge of tensed facts. He can do this either by revising the traditional definition of omniscience or else by maintaining that tense, while an objective feature of time, does not strictly belong to the factual content expressed by tensed sentences. Let us examine each strategy in turn.

The general problem with the strategy of revising the traditional definition of omniscience is that any adequate definition of a concept must be in line with our intuitive understanding of that concept. We are not free simply to "cook" the definition arbitrarily just to solve some problem under discussion. According to the traditional definition, a person is omniscient if and only if, for every fact, he knows that fact and does not believe its contradictory. On such a definition, if there are tensed facts, an omniscient person must know them. What plausible alternative definition of omniscience might the defender of divine timelessness offer?

Wierenga, as a sort of second line of defense, offers a revised account of omniscience which would not require God to know tensed facts.³³ Some facts, he says, are facts only from a particular perspective. They must be known to an omniscient being only if he shares that particular perspective. Thus, a person is omniscient if and only if, for every fact and every perspective, if something is a fact from a certain perspective, then that person must know that it is a fact from that perspective, and if that person shares that perspective, then he must know the fact in question. Wierenga treats moments of time as perspectives relative to which tensed facts exist. So while a temporal person existing on December 8, 1941, must (if he is omniscient) know the fact *Yesterday the Japanese attacked Pearl Harbor*, a timeless person must know only that from the perspective of December 8, 1941, it is a fact that *Yesterday the Japanese attacked Pearl Harbor*. On this definition God's omniscience does not require that He know the tensed fact, but only the tenseless fact that from a certain perspective a certain tensed fact exists.

Wierenga's revised definition of omniscience seems to me to be unacceptably "cooked." He is not denying that there are tensed facts. It might be tempting to understand his definition as an effort to eliminate tensed facts in favor of exclusively tenseless facts. For example, to say, "The Japanese attack is past relative to December 8, 1941," might sound like just a circumlocution for saying the attack is earlier than December 8, 1941, which is a tenseless fact. To say that something is past, present, or future relative to a time is just a misleading way of saying that it is earlier than, simultaneous with, or later than that time. One is not stating a tensed fact at all. If this were Wierenga's meaning, then he would simply be denying that there are tensed facts, and there would be no need to revise the definition of omniscience.

Rather, Wierenga wants to allow that there really are tensed facts but to maintain that an omniscient being need not know them. This claim seems quite implausible. On Wierenga's view temporal persons like you and me know an incalculable multitude of facts of which God is ignorant. Temporal persons know that the Japanese attack on Pearl Harbor is over; God has no idea whether it has occurred or not. He knows merely that for people on December 8, 1941, and thereafter, it is a fact that the attack is over. Since He does not know what time it actually is, He does not know any tensed facts. This is an unacceptably limited field of knowledge to qualify as omniscience.

Leftow also entertains the idea of revising the definition of omniscience in such a way that omniscience does not entail knowledge of all truths.³⁴

---

³³ Wierenga, *Nature of God*, 189.
³⁴ Leftow, *Time and Eternity*, 321-323.

Leftow's strategy here is strangely defeatist. He argues, in effect, that there are many sorts of truths which God cannot know, so there is no harm in admitting one more class of truths (namely, tensed truths) of which God is ignorant. I should have thought to the contrary that as Christian theologians we ought to construe God's knowledge as robustly as possible. If it turns out that there are truths God cannot know, that is no reason for further eroding the extent of His knowledge by denying Him knowledge of tensed truths! The rub is that Leftow is so deeply committed to divine timelessness that he is prepared to restrict or even jettison God's omniscience in order to preserve His timelessness. This strikes me as an odd set of theological priorities: abandoning a central doctrine that enjoys considerable scriptural support in order to hold on to a controversial doctrine at best intimated in Scripture.

In any case, does Leftow succeed in showing that there are truths which God cannot know? I think not. His examples of things God cannot know include how it feels to be oneself a failure or a sinner. But Leftow has confused *knowing how* with *knowing that*. Philosophers recognize that *knowing how* does not take *truths* as its object. God can know such truths as *Being a sinner feels lousy*, *Being a sinner feels depressing*, *Sinners feel guilty and hopeless*, and so on. These are the facts about how it feels to be oneself a sinner, and God knows these truths. When we talk about omniscience, we are speaking of knowledge in the sense of *knowing that*, where "knowing that" is followed by some truth. God's not knowing how it feels to be Himself a sinner is not an example of a truth He fails to know and so does not constitute a restriction on His omniscience. Leftow furnishes no example of any truth which might be conjoined with "knows that" such that we cannot say, "God knows that ___," where the blank is filled by the truth in question. Therefore, he has not adequately motivated our denying that knowledge of tensed truths properly belongs to omniscience.

It seems to me, therefore, that no adequate grounds have been given for thinking that someone could be omniscient and yet not know tensed truths. The traditional definition of omniscience requires it, and we have no grounds which do not involve special pleading for revising the usual definition.

So what about the second strategy for denying premise (3), namely, maintaining that tense does not, strictly speaking, belong to the factual content expressed by tensed sentences, even though tense is an objective feature of the world? This alternative takes us into very subtle issues in the philosophy of language. Although many philosophers think that the factual content expressed by tensed sentences includes tense, others construe the factual content tenselessly. The latter maintain that any indexical expressions in a sentence should be eliminated, along with the tense of the verb, in giving the factual content expressed by the sentence. For example, in a certain context of utterance, the sentence "I came by here yesterday" expresses the fact *Albert Wesselink comes by the Reformers' Wall in Geneva, Switzerland, on August 8, 1991.* In a different context that same sentence might express a wholly different fact. Tense could be analyzed as a feature of the mode in which the factual content is presented to someone expressing that content, or of the way in which a person grasps the factual content, or of the context of someone's believing the factual content. Alternatively, tense could be understood in terms of a person's ascribing to himself in a present-tense way the property of being such as the factual content expressed by the sentence specifies. On such analyses, an omniscient being could be timeless because tense is not part of the factual content of tensed sentences. Tense is real, all right, but since it does not belong to the factual content of a sentence, a being which knew only tenseless facts would on the traditional definition count as omniscient.

Fortunately, I do not think that a discussion of these recondite semantical theories will be necessary here, for even though I find such analyses plausible and attractive, I do not think that they ultimately serve to save the day for the defender of divine timelessness. For according to Christian theism, God is not merely factually omniscient, but also maximally excellent cognitively. For example, on the theories under discussion, a factually omniscient God would know such things as *God is omnipotent, God loves His creatures, God created the universe,* and so on. But He would not have to possess any first-person beliefs such as "I am omnipotent," "I love my creatures," "I created the universe," and so forth. God would not even have to know that He is God! A machine could count as omniscient under such analyses. But such a God or machine would clearly not have maximum cognitive excellence. In order to qualify as maximally excellent cognitively, God would have to entertain all and only the appropriate, true first-person beliefs about Himself. This would furnish Him with what philosophers call knowledge *de se* (first-person self-knowledge) in addition to mere knowledge *de re* (knowledge of a thing from a third-person perspective). Notice that in order to be maximally excellent cognitively, God would not have to possess all knowledge *de se* in the world, but only such knowledge *de se* as is appropriate to Himself. It would be a cognitive defect, not a perfection, for God to have the belief "I am Napoleon," though for Napoleon such a belief would be a perfection. The point is: Omniscience (on these theories) is not enough; God must be maximally excellent cognitively.

Now in the same way, it is a cognitive perfection to know what time it is, what is actually happening in the universe. A being whose knowledge is composed exclusively of tenseless facts is less excellent cognitively than a being who also knows what has occurred, what is occurring, and what will occur in the world. This latter person knows infinitely more than the former and is involved in no cognitive defect in so knowing. On the analogy of knowledge *de se,* we can refer to such knowledge as knowledge *de praesenti* (knowledge of the present). A being who lacks such knowledge is more ignorant and less excellent cognitively than a being who possesses it. Accordingly, if we adopt views according to which tense is extraneous to the factual content expressed by a tensed sentence, we should simply revise premise (3) to read

> 3'. If a temporal world exists, then if God is maximally excellent cognitively, then God has knowledge *de praesenti*

and, with appropriate revisions, the argument goes through as before.

In response to the contention that God, being maximally excellent cognitively, must have knowledge *de praesenti,* Kvanvig offers two alternatives.[35] One alternative would be to maintain that God grasps all moments of time directly. But this alternative makes no sense, for then all moments of time would be apprehended by God as present, not in the metaphorical sense of the eternal "present," but as temporally present. That would annul all temporal relations of *earlier than/later than* among events and leave God ignorant of which moment really is present. The other alternative is simply to give up the doctrine of divine timelessness. In Kvanvig's opinion the arguments for this doctrine are not compelling, so that if it is incompatible with God's omniscience or cognitive excellence, then it can be abandoned. This is the conclusion to which the present argument seems to be driving us.

Such a conclusion goes down hard with Leftow, however.[36] In his view, if timelessness and omniscience are incompatible, then we should give up the doctrine of omniscience. For he contends that a timeless God who is ignorant of tensed facts is more perfect on the whole than an omniscient God who is temporal.

How shall we assess such a comparison? We have already examined Leftow's claim that the life of a temporal being is inferior to the life of a timeless being due to the former's incompleteness.[37] While the argument has some

---

[35] Kvanvig, *All-Knowing God,* 159-160.
[36] Leftow, *Time and Eternity,* 323-326.
[37] See chapter 2, pages 67-74.

plausibility, we found its force diminished, interestingly, due to a temporal God's omniscient power to recall or anticipate past and future events. On the other hand, the superiority of temporal over timeless existence, we have discovered, lies in the ability which is afforded uniquely by temporal existence for God to be causally related to a temporal world. The incarnation of the second person of the Trinity is a special thorn in the side of advocates of divine timelessness. Leftow also attempts to downplay the importance of the attribute of omniscience, arguing that it is not essential to perfect knowledge. He rightly observes that cognitive perfection involves many other qualities than just the breadth of one's knowledge. Assuredly; but that is no reason to doubt that cognitive perfection should not encompass knowledge of tensed facts. Leftow also argues that omniscience is impossible, since God could not know the factual content expressed by sentences containing personal indexical words such as "I am overweight." But we have already seen how such knowledge *de se* can be handled without recourse to private, first-person facts. And in any case, placing one restriction on God's knowledge hardly makes it a matter of indifference if additional abridgments are proposed. To be as knowledgeable as possible is an important perfection which magnifies God's greatness. Thus, it seems to me a poor bargain, indeed, to auction off omniscience in order to purchase timelessness.

The attempt to deny premise (3) of the present argument thus seems to fare no better than the effort to refute premise (4). If God is omniscient, then given the existence of a temporal world, He cannot be ignorant of tensed facts.

From the premises of the argument, it follows that God is not timeless, which is to say, He is temporal. So in addition to the argument from God's real relation to the world, we now have a second powerful reason based on God's changing knowledge of tensed facts for thinking that God is in time.

Conclusion

On the basis of our foregoing discussion, we have seen comparatively weak grounds for affirming divine timelessness but two powerful arguments in favor of divine temporality. It would seem, then, that we should conclude that God is temporal.

But such a conclusion would be premature. For there does remain one way of escape still open for defenders of divine timelessness. The argument based on God's real relation to the world assumed the objective reality of temporal becoming, and the argument based on God's knowledge of the temporal world assumed the objective reality of tensed facts. If one denies the objective reality of temporal becoming and tensed facts, then the arguments are undercut. For in that case, nothing to which God is related ever comes into or passes out of being, and all facts tenselessly exist, so that God undergoes neither extrinsic nor intrinsic change. He can be the immutable, omniscient Sustainer and Knower of all things and, hence, exist timelessly.

In short, the defender of divine timelessness can escape the arguments of this chapter by embracing the static (or tenseless) theory of time.[38] According to that theory of time, all things and events in time are equally existent. Events in time are related by the tenseless relations of *earlier than, simultaneous with,* and *later than*. But the distinction between past, present, and future is not an objective distinction, being merely a subjective feature of consciousness. If there were no minds, there would be no past, present, or future. There would be just the four-dimensional space-time universe existing *en bloc*.

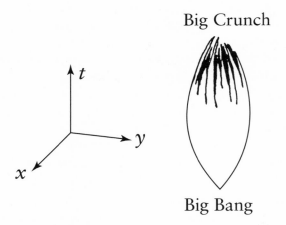

Fig. 3.1. A picture of a tenselessly existing space-time universe. The vertical dimension *t* represents time, which begins at the Big Bang and ends at the Big Crunch. The *x, y* dimensions represent three-dimensional space, one dimension of which cannot be pictured on the diagram because its place is assumed by the *t* dimension.

The reader looking at Fig. 3.1 represents God, who transcends space and time (of course, the reader must imagine himself as unchanging). The space-time universe is *intrinsically* temporal in that it has an internal dimension which, in virtue of its ordering relations (*earlier/later than*), is time. But the space-

---

[38] Recall the distinction drawn in chapter 2 (pages 69-70) between the two theories of time.

time universe is *extrinsically* timeless in that it is not embedded in some higher dimension (a hyper-space-time) but co-exists timelessly with God.

Given such a static view of time, it is easy to see why God never experiences extrinsic change in relation to temporal events. For there is no temporal becoming. Nothing in the space-time block ever comes into or goes out of being, nor does the space-time block as a whole come into being or pass away. It simply exists timelessly along with God. God is the Creator of the universe in the sense that the whole block and everything in it depends upon God for its existence. God by a single timeless act makes it exist. By the same act He causes all events to happen and things to exist at their tenseless temporal locations. Thus, God never acquires or loses any of His relations. God does not come into the relation *Creator of* with the Big Bang at t_0 and cease to stand in this relation at a later time t_1. Rather He timelessly stands in the *Creator of* relation to all events at their respective times. Thus, by denying the reality of temporal becoming, the defender of divine timelessness can undercut premise (2) of the argument for divine temporality based on God's real relation to the world. God is really related to the world, but He is not temporal.

Similarly, on the static theory of time there really are no tensed facts. The factual content of sentences containing tensed verbs and temporal indexicals includes only tenseless dates and tenseless relations of events. Linguistic tense is an egocentric feature of language users. It serves only to express the subjective perspective of the user. Thus, there really are no tensed facts for God to know. There is no objective truth about what is now happening in the universe, for "now" serves merely to signify the time relative to which the tensed judgment is made. Every person at every time in the space-time universe regards his time as "now" and others as "past" or "future." But in objective reality there is no "now" in the world. Everything just exists tenselessly. God, in knowing the tenseless times at which events occur and the tenseless temporal relations among them, knows all the objective facts there are. Thus, premise (3) of the argument for divine temporality based on God's knowledge of the temporal world is undercut. In knowing all tenseless facts, God is truly and timelessly omniscient.

The defender of divine timelessness therefore has a way out: He can adopt a static theory of time and deny the reality of tensed facts and temporal becoming. It is noteworthy, however, that almost no defender of divine timelessness has taken this route. Virtually the only proponent of timeless eternity to do so is Paul Helm. On his view there is no difference in the reality of the past, present, and future: "Do the times which are at present future to us exist, or not?" he asks. "Answer: they exist for God ... and they exist for those creatures contemporaneous with that future moment, for that moment is present to them, but it is not now present to us."[39] In the same way, "the past event ... belongs in its own time, and is therefore real, belonging to the ordered series of times which comprise the creation and which are ... eternally present to God."[40] Thus, Helm affirms, "in creation God brings into being (timelessly) the whole temporal matrix," and "God knows *at a glance* the whole of his temporally ordered creation...."[41] Similarly, tense is but an ephemeral feature of language; the truth conditions of tensed sentences are given by tenseless facts, facts that are known to God.[42] Helm thus appears to be one of the few advocates of divine timelessness who have seen and taken the way out.

If our discussion of the nature of divine eternity is not to end at this point, we have no option but to explore the viability of this escape route. Is the tenseless theory of time as credible as the tensed theory of time? In raising this question, we enter into the very heart of the philosophy of time and space. This is difficult and mysterious territory. One eminent metaphysician has called the nature of time "the most puzzling and paradoxical feature of the world."[43] But we have no choice: If we are to understand eternity, we must first understand time.

Recommended Reading

THE IMPOSSIBILITY OF ATEMPORAL PERSONHOOD

Yates, John C. *The Timelessness of God,* chapter 6. Lanham, Md.: University Press of America, 1990.

Leftow, Brian. *Time and Eternity,* chapter 13. Cornell Studies in Philosophy of Religion. Ithaca, N.Y.: Cornell University Press, 1991.

Walker, Ralph C. S. *Kant,* pp. 34-41. The Arguments of the Philosophers. London: Routledge & Kegan Paul, 1978.

DIVINE RELATIONS WITH THE WORLD

Craig, William Lane. "The Tensed vs. Tenseless Theory of Time: A Watershed for the Conception of Divine Eternity." In *Problems of Time and Tense,* pp. 221-250. Edited by Robin Le Poidevin. Oxford: Oxford University Press, 1998.

---

[39] Paul Helm, "Eternal Creation: The Doctrine of the Two Standpoints," in *The Doctrine of Creation,* ed. Colin Gunton (Edinburgh: T. & T. Clark, 1997), 42.
[40] Ibid., 43.
[41] Helm, *Eternal God,* 27, 26.
[42] Ibid., 25, 44, 47, 52, 79.
[43] E. J. Lowe, *The Possibility of Metaphysics* (Oxford: Clarendon, 1998), 84.

Padgett, Alan. *Eternity and the Nature of Time,* chapter 4. New York: St. Martin's, 1992.

Hill, William. "Does the World Make a Difference to God?" *Thomist* 38 (1974): 146-164.

Alston, William P. "Hartshorne and Aquinas: A Via Media." In *Existence and Actuality,* pp. 78-98. Edited by John B. Cobb, Jr., and Franklin I. Gamwell. Chicago: University of Chicago Press, 1984.

Craig, William Lane. "The Special Theory of Relativity and Theories of Divine Eternity." *Faith and Philosophy* 11 (1994): 19-37.

DIVINE KNOWLEDGE OF TENSED FACTS

Kretzmann, Norman. "Omniscience and Immutability." *Journal of Philosophy* 63 (1966): 409-421.

Castañeda, Hector-Neri. "Omniscience and Indexical Reference." *Journal of Philosophy* 64 (1967): 203-210.

Smith, Quentin. *Language and Time,* pp. 58-60. New York: Oxford University Press, 1993.

Davis, Stephen T. *Logic and the Nature of God,* chapters 2, 3. Grand Rapids, Mich.: Eerdmans, 1983.

Kvanvig, Jonathan L. *The Possibility of an All-Knowing God,* pp. 66-70, 154-159. New York: St. Martin's, 1986.

Wierenga, Edward R. *The Nature of God: An Inquiry into Divine Attributes,* pp. 179-185. Cornell Studies in the Philosophy of Religion. Ithaca, N.Y.: Cornell University Press, 1989.

Nelson, Herbert J. "Time(s), Eternity, and Duration." *International Journal for Philosophy of Religion* 22 (1987): 3-19.

Gale, Richard M. "Omniscience-Immutability Arguments." *American Philosophical Quarterly* 23 (1986): 319-335.

4

THE DYNAMIC CONCEPTION OF TIME

WE NOW ENTER UPON what one prominent philosopher has recently called "the most fundamental question in the philosophy of time": "Whether a static or a dynamic conception of the world is correct."[1] This question is not only fundamental for the philosophy of time but is also, as we have seen, foundational for our conception of divine eternity. For if the dynamic conception of time is correct, God is most plausibly understood to be temporal. What arguments, then, are there for and against the dynamic conception of time?

I. Arguments for a Dynamic Conception

1. *The Ineliminability of Tense*

EXPOSITION

We have already had occasion to mention tense as it plays a role in language. Although there are languages that do not express tenses by means of inflecting verbs, there is no tenseless language in the world.[2] This is doubtlessly due to the fact that language reflects our experience of the world as being tensed, as having a past, present, and future. Some philosophers, however, see an even deeper significance to linguistic tense.[3] They argue that linguistic tense is, as it were, a window on the world: Our language is tensed because reality is tensed. That is to say, there really are tensed facts which are objective features

---

[1] Michael Tooley, *Time, Tense, and Causation* (Oxford: Clarendon, 1997), 13.
[2] According to Gorman and Wessman, "all four thousand or so known languages enable their speakers to designate temporal relationships and to distinguish between past, present, and future events—though with varying degrees of difficulty" (Bernard S. Gorman and Alden E. Wessman, "The Emergence of Human Awareness and Concepts of Time," in *The Personal Experience of Time,* ed. Bernard S. Gorman and Alden E. Wessman [New York: Plenum Press, 1977], 44-45).
[3] See especially Richard M. Gale, *The Language of Time,* International Library of Philosophy and Scientific Method (London: Routledge & Kegan Paul, 1968); and Quentin Smith, *Language and Time* (New York: Oxford University Press, 1993).

of the world. For example, it is a fact that *Napoleon lost the Battle of Waterloo* and that *Bill Clinton will not be elected president in 2008*. Linguistic tense merely exhibits the tense that is a feature of time itself.

How might this be shown? Advocates of a dynamic conception of time argue that the ineliminability of tense from language and its indispensability for human life make it plausible that tense is a feature not merely of language but also of the world. Against this argument, defenders of a static view of time have pursued two strategies: They either try to show that tense can be eliminated from language without any loss of meaning or else they admit that tense cannot be eliminated from language but deny that this has any significance, since all one needs in order to make tensed sentences true or false is tenseless facts. Accordingly, the advocate of the dynamic view needs to thwart both of these strategies if his argument is to succeed.

We can formulate the linguistic argument for tensed facts as follows:

1. Tensed sentences apparently express tensed facts.

2. The apparent expression of tensed facts by tensed sentences should be accepted as correct unless

> i. tensed sentences are shown to be translatable into tenseless sentences without any loss of meaning

or

> ii. tensed facts are shown to be unnecessary for the truth of tensed sentences.

3. Tensed sentences have not been shown to be translatable into tenseless sentences without any loss of meaning.

4. Tensed facts have not been shown to be unnecessary for the truth of tensed sentences.

5. Therefore, the apparent expression of tensed facts by tensed sentences should be accepted as correct.

The argument purports to show that there are tensed facts about the world and that therefore the dynamic conception of time is correct.

CRITIQUE

Premise (1) of the argument is obviously true. By our tensed sentences we try to convey facts about the world. For example, when we say, "Churchill was the British Prime Minister during World War II," we are purporting to relay some fact about the world. Premise (2) is based on the conviction that unless we have some good reason to doubt this expression of tensed facts, we ought to accept it. Such a conviction seems quite reasonable. A correspondence view of truth holds that if a tensed sentence or statement is true, then it corresponds with reality. Therefore, if any tensed sentences are true, then reality must be tensed. There are only two known ways of escaping this conclusion, which are specified in clauses (i) and (ii). The crucial premises in the argument are therefore (3) and (4), which try to bar these routes. The route referred to in (3) has been called "The Old Tenseless Theory of Language" and the route mentioned in (4) "The New Tenseless Theory of Language." Let us examine each in turn.

The Old Tenseless Theory of Language

The Old Tenseless Theory seems to have originated with Bertrand Russell during the first decade of the twentieth century. Thereafter, for around three-quarters of a century, the standard answer of defenders of static time to the linguistic argument was that tense is a superfluous and even annoying feature of ordinary language which philosophically and scientifically trained minds are only too happy to discard.

De-tensers, as they are sometimes called, held that any tensed sentence can be translated without loss of meaning into a standard tenseless form. This was done in one of two ways. First, one could replace tensed expressions with tenseless verbs and dates/clock-times. For example, the tensed sentence "Mrs. Brown was not at home" could be translated into the tenseless sentence "Mrs. Brown *is* not at home on May 8, 1906" (the italics indicating that the "is" is tenseless).

Second, alternatively, one could replace tensed expressions with tenseless verbs and what are called "token-reflexive" expressions.[4] For example, "Mrs. Brown was not at home" could be translated into the tenseless, token-reflexive sentence, "Mrs. Brown *is* not at home earlier than this utterance." De-

---

[4] In the present context a "token" is a particular example of a type of thing. A sentence token is thus a particular utterance or inscription of a sentence. For example, when two people say, "Mrs. Brown is not at home," there are two tokens of the sentence. There is one sentence type, and in that sense they utter the same sentence; but there are two tokens of that sentence type, one spoken by the first individual and the other by the second. A sentence is token-reflexive if a token of that sentence refers to itself, e.g., "This sentence has five words."

tensers held that these tenseless translations have the same meaning as their tensed counterparts. Since such translations state merely tenseless facts, it follows that tense is a superfluous feature of ordinary language which gives us no insight into the nature of the world.

As the reader might surmise from its name, the Old Tenseless Theory of Language is now widely recognized as a failed project. Recent work in the philosophy of language has made it quite evident that the purported tenseless translations do not have the same meaning as their tensed counterparts. Three broad considerations undergird this judgment.

First, tensed sentences are informative in a way that their purported tenseless translations are not. Human thought and action would be paralyzed if the content of our beliefs were exclusively tenseless in character. The work of the Stanford philosopher John Perry has especially served to emphasize this point.[5] Perry invites us to imagine a faculty professor who holds the tenseless belief, "The faculty meeting *starts* at noon." All morning long the professor has held that belief, and therefore that belief cannot be the explanation why at noon he gets up and goes to the meeting. What explains the change in his behavior is that he has come to hold the tensed belief, "The meeting is starting now." As Perry notes, "These indexicals are essential, in that replacement of them by other terms destroys the force of the explanation...."[6] The tenseless counterparts of tensed sentences cannot motivate timely thought and action because they give us no knowledge *de praesenti* (knowledge of what is now the case).

This point is underlined by the ineptness of some of the supposed tenseless translations of tensed sentences. Take, for example, the tensed sentence "It is now 4:30." We can imagine situations in which a person's life would depend on his holding such a belief. But the tenseless counterpart of this sentence is either "It is 4:30 at 4:30," which is a mere tautology, or "It is 4:30 simultaneous with this utterance," which is useless unless we also know that "This utterance is occurring now," which is a tensed belief. In both cases the tenseless versions are insufficient to motivate timely action because they do not inform us whether or not it actually is 4:30.

The fact that tensed beliefs can motivate timely behavior in a way that the tenseless counterparts of those beliefs cannot is a convincing demonstration that tenseless sentences do not have the same meaning as the tensed sentences allegedly translated by them.

---

[5] John Perry, "The Problem of the Essential Indexical," *Noûs* 13 (1979): 3-29; idem, "Frege on Demonstratives," *Philosophical Review* 86 (1977): 474-477.
[6] Perry, "Essential Indexical," 4.

Second, tenseless date-sentences are informative in a way that their tensed counterparts are not. I might hold a tensed belief, for example, that "Mrs. Brown was not at home" without knowing or believing that she is not at home on May 8, 1906, as is stated in the tenseless version of the sentence. Perry also points out that if one has lost track of time, he might rationally believe that "The meeting *starts* at noon, September 16, 1976" and yet deny at that time that "The meeting is starting now." This shows that these two sentences cannot have the same meaning, as the Old Tenseless Theory asserts. Because tenseless date-sentences contain information which their tensed counterparts do not, it is evident that they do not have the same meaning.

Third, tensed sentences do not imply the existence of sentence tokens as do their token-reflexive counterparts. Consider the sentence, "No sentence tokens exist." This sentence is false, but it seems possible for it to be true (for example, during the Jurassic Period). But its tenseless translation is "No sentence tokens *exist* simultaneous with this utterance," which is a self-contradiction and therefore not even possibly true. Therefore, these sentences cannot have the same meaning. In general, anyone with a mastery of English understands that tensed sentences do not imply the existence of tokens of those sentences. It is not part of the meaning of a tensed sentence that it refer to itself.

For all these reasons the Old Tenseless Theory of Language has been universally abandoned by defenders of the static view of time. Linguistic tense is ineliminable. Premise (3) of the argument is therefore no longer contested by static time theorists.

The New Tenseless Theory of Language

Having retreated from the translatability claim, de-tensers have recently regrouped behind the New Tenseless Theory of Language, which has been brilliantly formulated and defended by the Cambridge philosopher D. H. Mellor.[7] Mellor is quite insistent on the fact that tensed sentences cannot be translated into tenseless sentences. But he does think that tensed sentences can be given tenseless truth conditions. To give the truth conditions of a sentence is to state the conditions under which the sentence is true. Mellor holds that a present-tense sentence token is true if and only if that token occurs at the time of the event it describes; a past-tense sentence token is true if and only if that token occurs at a specified time later than the event described; and a future-tense token is true if and only if the token occurs at a specified time

---

[7] D. H. Mellor, *Real Time* (Cambridge: Cambridge University Press, 1981).

earlier than the relevant event. So, for example, take the sentence "It is now 1980" and call some token of this sentence "*S.*" *S* is true if and only if *S occurs* in 1980. "It is now 1980" obviously does not mean "*S occurs* in 1980." Nevertheless, "*S occurs* in 1980" does state the tenseless truth conditions for *S.* That is to say, a token of the sentence "It is now 1980" is true just in case that token is uttered in 1980. At any other time it would be false. But if it occurs in 1980, that token is true.

Now Mellor takes the tenseless truth conditions to be what makes the tensed sentences true. Since those conditions are tenseless, no tensed facts are necessary to make tensed sentences true. All that is needed is tenseless facts.

Now all this still does not serve to show that tensed facts are dispensable, as the static time theorist must show if he is to undercut premise (4). Mellor recognizes this, and he goes on to argue that the tenseless sentences stating the truth conditions of tensed sentences also give us the rules governing when we should use those tensed tokens. For example, anyone who knows the truth conditions of "It is now 1980" will know when to use a token of that sentence, namely, only during 1980. But how, we may wonder, can we know when to use such a token unless we can grasp the tensed fact that it is now 1980? We all have a certain now-awareness, which Mellor calls the *presence* (or better, *presentness*) of experience, which must be conjoined to any rules for using tensed sentence tokens if we are to use them successfully. Indeed, Mellor recognizes that "This curious phenomenon, the experienced temporal presence of experience, is the crux of the tensed view of time, and the tenseless camp must somehow explain it away. . . . Without a tenseless account of it, tenseless truth conditions on their own will never dispose of tensed facts."[8] Mellor therefore argues that although we observe our experience to be present, it in fact is not.

So as not to bite off more than we can chew, let us set to the side for the time being Mellor's denial of the presentness of experience and focus on his claim that only tenseless facts are required to make tensed sentences true.[9] There are a number of reasons to think that Mellor's New Tenseless Theory of Language is no more successful than the Old Tenseless Theory in rendering tensed facts superfluous.

First, the New Tenseless Theory violates the laws of logic. Because Mellor is offering truth conditions for sentence tokens, not types, two different tokens of the same sentence type must have different truth conditions. For

---

[8] Ibid., 6, 50.
[9] We shall come back to the presentness of experience in the next section.

example, imagine two people simultaneously saying, "It is now 1980." Call one token of this sentence R and the other S. On Mellor's analysis the truth conditions of R and S are as follows:[10]

(R) "It is now 1980" ≡ *R occurs* in 1980

(S) "It is now 1980" ≡ *S occurs* in 1980

The problem with this analysis is that while R necessarily implies S, "*R occurs* in 1980" does not necessarily imply "*S occurs* in 1980"— and yet R is said to be logically equivalent to "*R occurs* in 1980" and S logically equivalent to "*S occurs* in 1980"! Worse, if R is logically equivalent to S and S is logically equivalent to "*S occurs* in 1980," then R should be logically equivalent to "*S occurs* in 1980," which it obviously is not. Moreover, if the only facts stated by R and S are the tenseless facts constituting their respective truth conditions, then since these truth conditions are not the same fact, R and S do not state the same fact, which is absurd. Thus, Mellor's account of the truth conditions of tensed sentences is logically incoherent.

All of this implies that Mellor has failed to state properly the truth conditions of tensed sentences. Rather what he has given us is a tenseless rule for when a token can be truly uttered:

R *is* truly uttered ≡ *R occurs* in 1980

S *is* truly uttered ≡ *S occurs* in 1980

Such a tenseless rule of use is unproblematic; but it does nothing to suggest that R or S themselves have tenseless truth conditions.

The advocate of the dynamic theory of time may offer instead tensed truth conditions of tensed sentences. In order to do so, one must challenge the assumption that when it is said, "R is true if and only if . . ." the "is" is tenseless. If the "is" is present tense, then we shall have truth conditions for R's being presently true or true now. When we state tensed truth conditions, we find that we are employing the standard schema for truth $Tp \equiv p$.[11] Thus

---

[10] The symbol "≡" means "if and only if." This signifies the logical equivalence of the statements connected by the symbol; that is to say, the statements mutually imply each other.

[11] This is called the Tarski schema for truth after the logician Alfred Tarski. It means that a statement is true if and only if what that statement says is the case. Thus, it is true that "Snow is white" if and only if snow is white.

"The Battle of Waterloo occurred in 1815" is (presently) true if and only if the Battle of Waterloo occurred in 1815. This just is a view of truth as correspondence with reality. What all this implies is that there are tensed states of affairs that are actual now, or, in other words, that there are tensed facts.

Second, the New Tenseless Theory offers no coherent account of tensed sentences which are never tokened. This problem is a holdover from the Old Tenseless Theory. What truth conditions are to be given for the sentence "There are no sentence tokens"? On Mellor's analysis such a sentence cannot possibly be true; but it seems obvious that everyone could be silent for a minute!

Mellor's colleague Jeremy Butterfield has an acute awareness of the problem this poses for the New Tenseless Theory.[12] He is embarrassed by the fact that on such an analysis something not talked about during its existence is never present, since there are no present-tense truths about it. Equally awkward is the fact that this thing could be future or past, if people predicted or recalled its existence, even though it is never present. Something else which was talked about during its existence but was unanticipated and immediately forgotten is present while it exists, but never future or past. Butterfield tries to remedy this defect by allowing untokened sentences to be true or to express truths. But he will not admit tensed facts. His view seems to be that tensed sentences are true at some times and false at others and that tenseless truth conditions can be given for a sentence which states at what times a tensed sentence is true. For some tensed sentence *S* we can formulate a tenseless sentence *T* stating precisely when *S* is true. Thus Butterfield's account only gives us truth conditions for tenseless sentences such as "*S is* true in the Jurassic Period," but not truth conditions for *S* itself. If *S* itself is true, though untokened, only during the Jurassic Period, then *S* states a tensed fact, and Butterfield has failed to provide tenseless truth conditions for it. Thus, in attempting to remedy the defective token-reflexive analysis of the New Theory, Butterfield has inadvertently backed himself into affirming the reality of tensed facts.

Third, the New Tenseless Theory confuses truth conditions with truth makers of tensed sentences. Protagonists of the New Tenseless Theory of Language consistently take the fact stated by a tensed sentence's tenseless truth conditions to be what *makes* the tensed sentence true. But such an assumption is based on a confusion. Stating truth conditions is a semantic

---

[12] Jeremy Butterfield, "Indexicals and Tense," in *Exercises in Analysis,* ed. Ian Hacking (Cambridge: Cambridge University Press, 1985), 69-87.

exercise aimed at specifying the conditions under which a sentence has a determinate truth value. But there is no reason to think that what serves as a sentence's truth conditions is also what makes the sentence true. One could lay down adequate truth conditions for any tensed sentence S by stating, for example, that S is true if and only if God believes what S states. But God's believing what S states is not what makes S true; indeed, quite the opposite: God believes what S states because S is true.

Now Mellor might respond by contending that truth conditions which also provide a rule for using tensed sentences do give us the facts which make those sentences true. But I think there are good counter-examples to this claim. Consider, for example, the truth conditions typically given for modal sentences—sentences about what is necessary or possible. According to the usual account, a statement is necessarily true if and only if it is true in all logically possible worlds, and a statement is possibly true if and only if it is true in some possible world. Given these truth conditions, I also understand how to use such modal sentences. But the semantics of possible worlds tells me nothing about what makes modal sentences true. As Alvin Plantinga, whose book *The Nature of Necessity* has become a classic in this area, explains,

> we can't sensibly *explain* necessity as *truth in all possible worlds;* nor can we say that *p*'s being true in all possible worlds is what *makes p* necessary. It may still be extremely useful to note the equivalence of *p is necessary* and *p is true in all possible worlds*: it is useful in the way diagrams and definitions are in mathematics; it enables us to see connections, entertain propositions and resolve questions that could otherwise be seen, entertained and resolved only with the greatest difficulty if at all.[13]

Thus the semantics of possible worlds constitutes a clear counter-example to the New Tenseless Theory's assumption that truth conditions which provide a rule for using tensed sentences give us the facts which make those sentences true.

Consider as well the semantics for counterfactual sentences about what would be the case or might be the case, if something else were the case. According to the usual analysis, a sentence about what would be the case is true if and only if in all possible worlds most similar to the actual world in which the antecedent clause of the counterfactual sentence is true, the consequent clause is also true. Thus the sentence "If Buchanan had won the

---

[13] Alvin Plantinga, "Reply to Robert Adams," in *Alvin Plantinga,* ed. James Tomberlin and Peter van Inwagen, Profiles 5 (Dordrecht: D. Reidel, 1985), 378.

Republican nomination in 1992, he would have lost the election" is true if and only if in all possible worlds most similar to the actual world in which Buchanan wins the nomination, he loses the election. A sentence about what might be the case is true if and only if the consequent clause is true in some of the most similar worlds in which the antecedent is true. These truth conditions also give us the rule for how to use such counterfactual statements. But once again, they do nothing to explain what makes the counterfactual statement true. Plantinga remarks,

> ... we can't look to similarity, among possible worlds, as *explaining counterfactuality, or as founding* or *grounding* it. (Indeed, any founding or grounding in the neighborhood goes in the opposite direction.) We can't say that the truth of A → C is *explained* by the relevant statements about possible worlds, or that the relevant similarity relation is what *makes* it true.[14]

Wholly apart from these counter-examples, the deeper philosophical issue here is the difference between conditions and grounds. Truth conditions are purely logical conditions and are not meant to constitute grounds for something else. For example, "Socrates died" and "Xanthippe [Socrates's wife] became a widow" are logically equivalent, but it would be completely wrong-headed to say that the former is made true by the latter or that the latter is the ground of the former. Thus, stating truth conditions is not the same thing as stating the grounds of truth. Even if the New Tenseless Theory of Language were correct in its statement of the truth conditions of tensed sentences, I see no reason to think that these disclose to us the facts which make the sentences true. On the contrary, if tensed sentences have truth makers, it seems plausible that it is the tensed facts expressed by tensed sentences which make them true, regardless of what truth conditions might be offered for them.

Remarkably, in the recently revised version of his book, Mellor himself abandons the New Tenseless Theory of Language under the force of objections such as the above.[15] But Mellor is not ready to admit the existence of tensed facts; instead he proposes another theory, what he calls the Indexical Tenseless Theory, to replace the untenable New Tenseless Theory. In proposing this new theory, Mellor makes it clear that his interest is not so much in truth *conditions* of tensed sentences as in their truth *makers*. Mellor is actually prepared to admit that there are tensed facts corresponding to what

---

[14] Ibid.
[15] D. H. Mellor, *Real Time II* (London: Routledge, 1998), xi, 32.

tensed sentences report. For example, corresponding to the tensed sentence "Jim races tomorrow" is the tensed fact that Jim races tomorrow.[16] This is a striking concession on Mellor's part, for at face value it grants precisely what defenders of the tensed theory of time have been saying. But Mellor maintains that the concession is merely apparent, for he insists that these facts are not what makes the tensed sentences true. What makes them true are exclusively tenseless facts, and so the argument for a tensed theory of time fails.

The Achilles' heel of the New Tenseless Theory was its reliance on sentence tokens as the bearers of truth. So what will Mellor substitute in their place as the truth bearers in his Indexical Theory? The answer is not at all clear, but it seems that sentence types rather than sentence tokens are the primary truth bearers. This gets around the problem of how there can be truths such as "There are no sentence tokens." For even if no token of a sentence type ever exists, the sentence type itself is a kind of abstract entity that exists regardless. So what makes tensed sentence types true if it is not the tensed facts which they report?

Mellor claims that any tensed sentence type S about an event E is made true at any time t by the fact that t precedes (or follows) E by the same amount of time as S says the present precedes (or follows) E.[17] So, for example, if $S = $ *Jim will race tomorrow,* then what makes S true on, say, June 1, 1999, is the fact that June 1, 1999 is one day before Jim's race. But that fact—*June 1, 1999 is one day before Jim's race*—is a tenseless fact. Thus, tensed facts are not needed to make tensed sentences true. So the ineliminability of tense from language does not require the existence of tensed facts in the robust sense of truth makers.

Mellor's theory just assumes that tensed sentences have truth makers, which is a controversial assumption.[18] But let that pass. Even granted that assumption, Mellor's theory does nothing to show that tensed sentences do not have tensed truth makers as well as tenseless truth makers. Truth-maker theorists universally acknowledge that there is no one-to-one correspondence between truths and truth makers—some truths may have multiple truth makers. So the question is whether the tensed facts which Mellor now admits exist might not be truth makers. In order to defeat the argument from the ineliminability of tense, Mellor must show that tensed facts are not truth makers.

---

[16] Ibid., 25.
[17] Ibid., 34.
[18] A view of truth as correspondence does not imply the existence of truth makers, and only a minority of philosophers explicitly endorse the idea that there are such entities as truth makers. Moreover, even among truth-maker theorists some hold that true past- and future-tense sentences do not have truth makers—rather the present-tense counterparts of such sentences did or will have truth makers.

For even the demonstration that tenseless facts are truth makers of tensed sentences does not prove that tensed facts are not their truth makers as well. If tensed facts exist, then it is very difficult to see why the sentences corresponding to them would not be true in virtue of such facts. Since he now concedes the existence of tensed facts, Mellor must show them to be so effete that tensed sentences are not, so to speak, overdetermined by the facts, rendered true by both tensed facts and tenseless facts. That he has not done.

But Mellor's revised theory has even deeper problems than that. The strength of the New Tenseless Theory was that it claimed to state the truth-conditions—and even, in Mellor's mind, the truth makers—of any tensed sentence, period. But Mellor's revised theory does not attempt to do either one. We are left wondering as to the truth conditions of *S*. Mellor's revised account cannot be construed in truth conditional terms, for then we should have:

S is true at $t \equiv t$ precedes (or follows) *E* by the same amount of time as *S* says the present precedes (or follows) *E*

Here we have truth conditions, not for *S* but for "*S* is true at *t*." These are truth conditions of a tenseless sentence, not of a tensed sentence! Considered as giving us *S*'s truth maker, the indexed account again fails to tell us what makes *S* true—it only tells us what supposedly makes *S* true at *t*. But as we saw in our discussion of tensed truth conditions, we want to know what makes (present tense) *S* true, period. We want to know, not what makes *Jim races tomorrow* true on June 1, but what makes it true that *Jim races tomorrow* or that *Jim is racing*. If tensed sentence types need truth makers, then it is tensed facts which are the truth makers of such tensed sentence types. For if there are no tensed truth makers, then it is inexplicable why *S* is true—not true at *t*, mind you, but simply true.

Defenders of the tenseless view of time thus appear to have failed in their attempt to undermine the reality of tensed facts. Neither the Old Tenseless Theory of Language, the New Tenseless Theory, nor Mellor's most recent Indexical Tenseless Theory succeeds in sloughing off the reality of tensed facts. On the contrary, it seems plausible that an adequate treatment of the truth conditions or truth makers of tensed sentences requires the reality of tensed facts.

But the defender of static time has one last card to play: De-tensers frequently claim that a parallel argument for the reality of spatial "tenses" can be constructed and that since spatial "tenses" obviously do not exist, the argument leading to such a conclusion must be fallacious. Since the arguments are entirely parallel, de-tensers conclude that it is fallacious to infer that either temporal or spatial tenses are objective.

De-tensers note that the spatial indexical "here" is entirely parallel to the temporal indexical "now." Spatial locations relative to "here," such as ten miles north of here or ten miles south of here, are analogous to future and past times. Moreover, some of the most memorable examples of "the essential indexical" concern spatial indexicals which serve to locate oneself. For example, someone lost in the stacks of the Stanford University library knows he is "here" but wants to know where "here" is. Similarly, someone looking at a map of the library can know that the circulation desk is on the second floor, but he will not try to check out his books unless he believes the circulation desk is "here." So spatial indexicals are every bit as ineliminable and indispensable as temporal indexicals. Yet no one believes that spatial "tenses" are real, that there is an objective "here" in the world independent of conscious beings. "Here" is just a subjective perspective on a world which exists in space independently of such perspectives. The same is true of "now" and temporal tenses.

This is a powerful rejoinder, but it seems to me that the theorist of dynamic time has the resources to meet it. One of the failings of this rejoinder is that it concerns indexical words only. But, as we have seen, tense in language is far from limited to indexical words alone. The dynamic time theorist could freely admit that indexical words express ego-centric perspectives. Just as we would not hold it to have been true at some time t during the Jurassic period at some location l on the North American continent that "A trachodon is laying her eggs here," so we would not hold it to have been true that "A trachodon is now laying her eggs." "Now" like "here" expresses the viewpoint of a conscious subject. But that does not imply that tense is subjective. For it was true at t, l that "A trachodon is laying her eggs," not merely that "A trachodon *lays* her eggs at t, l." The fact that indexicals are ego-centric does not imply that the present tense is unreal.

Secondly, the dynamic time theorist can eliminate spatial "tenses" by providing a reductive analysis of them. The first intimations of how to eliminate spatially perspectival facts may be found in Bertrand Russell's reflections on the "this-ness" of experience.[19] "This" and "that" are indexical words called demonstratives, which we use to designate one thing rather than another. Russell observed that instead of regarding "this-ness" as fundamental, we

---

[19] Bertrand Russell, *An Inquiry into Meaning and Truth* (London: George Allen & Unwin, 1940), 108-110.

could analyze it in terms of "I-now." For example, "this" is what I am now designating. In this case the "I-now" becomes fundamental and irreducible. The advocate of dynamic time can utilize this same insight to reductively analyze "here" in terms of the location of "I-now." "Here" is where I am now located. We can even trade in "now" for the simple present tense: "Here" is where I am located. On this analysis, spatial locations as given by coordinates (such as longitude and latitude) are objective, but spatial perspectives such as "here" and "there" are not. Because the "I-now" is irreducible, this analysis entails the objective existence of the self and the present. While such an analysis may not be congenial to reductive materialists, who want to get rid of the self, it is an analysis wholly in accord with Christian theism, which regards selves as genuine agents, God Himself being the paradigm example.

This analysis does raise a further difficulty, however. It seems to imply that there are private, first-person truths accessible only to individual selves, for example, that *I am Napoleon Bonaparte*. The only person who could grasp this private truth was Napoleon himself. Although a few philosophers hold that there are such truths, most do not. It seems bizarre to think that when someone says to me, "Tell Jan that I'm coming at 3:00," I cannot really grasp or communicate this fact, but can only tell Jan some other, quite different fact—for example, that Elaine is coming at 3:00. Moreover, if there are such private truths, then God is not omniscient, since He does not know, for instance, that He Himself is Napoleon. It is therefore desirable to find an account of the self and the present which does not commit us to the existence of private truths.

One alternative would be to hold that the factual content of sentences containing indexical words is indexical-free, "I" serving to express the individual essence of the speaker and "now" the present tense.[20] Thus, the factual content of the sentence "I am now eating lunch," when uttered in a specific context, would be something like *Jonathan Kvanvig is eating lunch* (letting the proper name stand in for Kvanvig's individual essence). When Kvanvig grasps this content, he expresses it with the first-person "I" because he is Kvanvig. But when we grasp this content we express it differently. Thus, we know the same truth, but because real, distinct selves exist, those selves grasp this truth differently. We could even go so far as to eliminate tense altogether from the factual content expressed by a tensed sentence, insisting, however, that because tense is real and we exist in the present, we grasp directly truths referring to the time which is present and indirectly truths referring to

---

[20] Compare Kvanvig's account discussed in chapter 3, pages 100-101.

other times. The self and the present are thus real even if they are not, as such, part of the factual content expressed by a sentence.

A second alternative would be to construe factual content in terms of a person's ascribing properties to himself.[21] Corresponding to the factual content expressed by a sentence one can conceive of a property of being in a world of that sort, which one ascribes to himself. For example, instead of holding that I believe a certain fact expressed by the sentence "David Lewis teaches at Princeton," we can hold that I ascribe to myself the property of *inhabiting a world in which David Lewis teaches at Princeton*. But when David Lewis asserts, "I teach at Princeton," he ascribes to himself the property of *teaching at Princeton*. Similarly, when one ascribes to oneself properties involving reference to the present time, one expresses this in the present tense or with indexicals such as "now." One need not hold that all sentences must be analyzed in terms of the self-ascription of properties; but one could hold that in the case of knowledge *de se* and *de praesenti*, at least, what one is doing is ascribing appropriate properties to oneself at the present time, not accessing private facts.

Either of these accounts seems viable, and no doubt others could be formulated as well, so we need not endorse one. The presence of such viable accounts suggests that one's commitment to the reality of the self and the present does not imply commitment to privately accessible facts. Therefore, the last objection on the part of de-tensers to the reality of tensed facts collapses.

In conclusion, it does seem to me that the linguistic argument for the objective reality of tense is a good one. Tensed sentences appear to express tensed facts, and neither the Old nor the New nor the Indexical Tenseless Theory of Language has been able to dispense with them. By positing the objective reality of the self and the present, the partisan of dynamic time can provide a reductive analysis of spatial indexicals which does not commit him to private facts. It therefore follows that unless better arguments can be marshaled in support of the static view of time, the reality of tensed facts and, accordingly, the dynamic theory of time should be accepted.

2. *Our Experience of Tense*

EXPOSITION

We have seen that tense in language is plausibly taken to be a reflection of tense in the world. But wholly apart from language, we experience the real-

---

[21] See David Lewis, "Attitudes *de dicto* and *de se*," *Philosophical Review* 88 (1979): 513-543; Roderick Chisholm, *The First Person* (Minneapolis: University of Minnesota Press, 1981); idem, "Why Singular Propositions?" in *Themes from Kaplan*, ed. Joseph Almog, John Perry, and Howard Wettstein (Oxford: Oxford University Press, 1989), 145-150.

ity of tense in a variety of ways that are so evident and so pervasive that the belief in the objective reality of past, present, and future is a universal feature of human experience. Here we move from the philosophy of language into the field of phenomenology, which seeks to provide a description of human experience.

Phenomenological analyses of temporal consciousness have emphasized the centrality of past, present, and future to our experience of time. In his classic analysis of temporal consciousness, the great phenomenologist Edmund Husserl described our experience of time in terms of remembering the past and anticipating the future, both anchored in consciousness of the "now." The transformation of a "now"-consciousness to a past-consciousness and its replacement by a new "now"-consciousness, says Husserl, "is part of the essence of time consciousness."[22]

Similarly, the psychologist William Friedman, who has made a career of the study of our consciousness of time, reports that "the division between past, present, and future so deeply permeates our experience that it is hard to imagine its absence."[23] He says that we have "an irresistible tendency to believe in a present. Most of us find quite startling the claim of some physicists and philosophers that the present has no special status in the physical world, that there is only a sequence of times, that the past, present, and future are only distinguishable in human consciousness."[24]

Consequently virtually all philosophers of time and space, even those who hold to a static view of time, admit that the view of the common man is that time involves a real distinction between past, present, and future. One advocate of the static view grumps that the dynamic understanding of time is so deeply ingrained in us that it seems "programmed by original sin"![25] The advocate of the dynamic view of time may plausibly contend that our experience of tense ought to be accepted as veridical, or trustworthy, unless we are given some more powerful reason for denying it.

The dynamic time theorist might formulate an argument to the effect that the objective reality of tense is the best explanation of our experience of tense. But it seems to me that our belief in the reality of tense is much more fundamental than such an argument suggests. We do not adopt the belief in an

---

[22] Edmund Husserl, *The Phenomenology of Internal Time-Consciousness,* ed. Martin Heidegger, trans. James Churchill, with an introduction by Calvin O. Schrag (Bloomington, Ind.: Indiana University Press, 1964), 86.

[23] William Friedman, *About Time* (Cambridge, Mass.: MIT Press, 1990), 92.

[24] Ibid., 2.

[25] J. J. C. Smart, "Spacetime and Individuals," in *Logic and Art,* ed. Richard Rudner and Israel Scheffler (Indianapolis: Bobbs-Merrill, 1972), 19-20.

objective difference between the past, present, and future in an attempt to *explain* our experience of the temporal world. Rather our belief in this case is what epistemologists call "a properly basic belief."[26]

A basic belief is a belief which is not believed on the basis of some underlying belief but is rather a foundational belief which we simply form in certain situations. For example, when I look out the window and form the belief "There is a tree," I am definitely *not* reasoning, "I am receiving certain sensory stimuli such that a tree is appearing to me. The best explanation of this sensory phenomenon is that there really is a tree and that therefore I am having this experience." Rather, in such a situation, I just automatically and immediately form the belief "There is a tree."

Now almost any belief could be held in a basic way by someone, but that does not mean that just any belief can be *properly* basic. In order to be properly basic, a belief must be grounded in the appropriate circumstances. Otherwise the belief is irrational. For example, if I look at my office wall and form the belief "There is a tree," then such a belief is not properly basic for me, since it is not grounded in appropriate circumstances. But if I am in the circumstances of looking at a tree, then such a belief is properly basic for me.

A little reflection discloses that the vast majority of our beliefs are properly basic beliefs. Perceptual beliefs, memory beliefs, and beliefs based on testimony are just some of the classes of beliefs which we hold to in a properly basic way. Properly basic beliefs can differ from one another with respect to how deeply ingrained they are and how strongly they are held. A deeply ingrained belief is one which, if abandoned, would force us to change many other beliefs as well. A strongly held belief is one which I hold very tenaciously, being unwilling to give it up lightly.

It is important to understand that properly basic beliefs are defeasible, that is to say, they can be shown to be false. For example, while visiting Disney World I might form the belief "There is a tree," which would be a properly basic belief for me in those circumstances, until I discover upon close inspection that the "tree" is a mere simulation. In such a case we say that my belief has been *defeated*. If I am to remain rational, I must now abandon the original belief that I saw a tree.

A belief's being properly basic thus implies that I am justified in holding to that belief unless and until it is defeated. We may say that such a belief is justified at face value (*prima facie*). For example, take the belief that "The

---

[26] For an account of this notion see Alvin Plantinga, "Reason and Belief in God," in *Faith and Rationality*, ed. Alvin Plantinga and Nicholas Wolterstorff (Notre Dame, Ind.: University of Notre Dame Press, 1983), 47-63.

external world is real." It is possible that you are really a brain in a vat of chemicals, being stimulated with electrodes by some mad scientist to believe that you are sitting there reading this book. Indeed, there is no way to prove this hypothesis wrong. But that does not imply that your belief in the reality of the external world is unjustified. On the contrary, it is a properly basic belief grounded in your experience and is as such justified until some defeater comes along. This belief is not defeated by the mere *possibility* that you are a brain in a vat. For there is no warrant for thinking that you are, in fact, a brain in a vat. Indeed, our belief in the reality of the external world is so deeply ingrained and strongly held that any successful defeater of this belief would have to possess enormous warrant. In the absence of any successful defeater, you are perfectly justified in taking your experience of the external world to be veridical.

Now the advocate of a dynamic view of time may argue similarly concerning our belief in the past, present, and future. Belief in the objective reality of tense is a properly basic belief which is universal among mankind. It therefore follows that anyone who denies this belief (and who is aware that he has no good defeaters of that belief) is irrational, for such a person fails to hold to a belief which is for him properly basic.

Sometimes advocates of a tenseless view of time assert that our experience of past, present, and future need not be taken as veridical, since we can imagine a universe exactly like this one which is a four-dimensional block universe containing individuals whose mental states correspond exactly to our mental states in this world. "But then surely our copies in the block universe would have the same experiences that we do—in which case they are not distinctive of a dynamic universe after all. Things would seem this way, even if we ourselves were elements of a block universe."[27] But this is like arguing that because a brain in a vat would have the same experiences of the external world that we do, therefore we no longer have any grounds for regarding our experiences as veridical! In the absence of some sort of defeater of beliefs grounded by such experiences, these experiences do provide warrant for those beliefs.

We can formulate this argument as follows:

1. Belief in the objective reality of the distinction between past, present, and future is properly basic.

---
[27] Huw Price, *Time's Arrow and Archimedes' Point* (New York: Oxford University Press, 1996), 15.

2. If our belief in the objective reality of the distinction between past, present, and future is properly basic, then we are *prima facie* justified in holding this belief.

3. Therefore, we are *prima facie* justified in holding our belief in the objective reality of the distinction between past, present, and future.

Since premise (2) is true by definition of "properly basic belief," the only disputable premise is (1).

Critique

The Presentness of Experience

Let us examine more closely our experience of time in order to evaluate premise (1). To begin with the most obvious, we experience events as present. Our belief that events are happening presently is really no different than our belief that they are happening—and this latter belief is a basic belief grounded in our perceptual experience.

D. H. Mellor, as a proponent of the static view of time, does not believe that there really is a present. Therefore, he says, we cannot, despite appearances, be experiencing it. Mellor thus goes to great lengths to explain away our experience of the present.

First, he argues that we do not really observe the tense of events.[28] He gives an illustration of observing astronomical events through a telescope. When I look at the stars, I seem to be observing the events as presently happening; but we know that they actually occurred millions of years ago. Thus, what I see is the *order* in which events occurred, but my observations do not tell me the tense of the events. Therefore, when we think that we are observing any event to be present, we are simply confused. We do not observe the event itself to be present; rather we observe our *experience* of the event to be present.

Now it seems to me that Mellor's objection is ineffective against the argument as we have framed it. For clearly I do not form a belief such as "The phone is ringing" by inferring it from a more foundational belief such as "My experience of the phone's ringing is present." Typically, I do not form any belief like the latter at all. My beliefs about the tense of events is not inferred, but basic. As for the illustration of events viewed through a telescope, all that proves is that my beliefs about the tense of events is defeasible and sometimes wrong. One might as well argue that perceptual beliefs are not properly basic

---

[28] Mellor, *Real Time*, 26.

because things viewed through a microscope are observed to be larger than they are! Just because our sense perceptions are sometimes mistaken is no reason to think that we do not perceive things. In the same way, mistaken observations of the presentness of certain events do not prove that we make no such observations.[29] In most cases, the events we observe fall within the limits of the specious present, so that our observations of events as present are veridical and our judgments to that effect properly basic.

In any case, Mellor admits that we do observe our experiences to be present. This is the so-called presentness of experience. Even if I can be mistaken about the presentness of a supernova observed through a telescope, I cannot be mistaken about the presentness of my experience of observing the supernova. If I observe my experiences to be present, am I not observing the tense of these mental events?

No, says Mellor, for "although we observe our experience to be present, it really isn't."[30] This is a paradoxical statement. Mellor admits that when I make the judgment that my experience is present, I cannot be mistaken. He writes,

> So judging my experience to be present is much like my judging it to be painless. On the one hand, the judgment is not one I have to make.... But on the other hand, if I do make it, I am bound to be right, just as when I judge my experience to be painless. The presence of experience ... is something of which one's awareness is infallible.
>
> ... No matter who I am or whenever I judge my experience to be present, that judgment will be true.[31]

But if my observation of the presentness of my experience is analogous to my observation of whether my experiences are painful, if I am *bound to be right* in judging my experience to be present, if my awareness of the presentness of my experience is *infallible,* if my judgment that my experience is present will be *true every time,* then how can it be the case that, as Mellor says, "it really isn't"? If my belief that "My experience of observing the supernova is present"

---

[29] If Mellor's argument were successful, it would also imply that we do not even observe events to be earlier or later than one another. For a common problem in astronomy is that a galaxy farther away than another can appear to be the same distance from us because the more distant one is bigger and therefore of the same brightness. Someone unaware of this difference in distance would think that the galactic events he observes in both are simultaneous, when in fact events in the more distant galaxy occurred earlier than the events observed in the other.

[30] Mellor, *Real Time,* 26.

[31] Ibid., 53.

is indefeasible, as Mellor admits, then how can that experience not be present, even if the supernova itself is not?

Mellor's answer is that, while the belief that one's experience is present may have important cognitive significance, nonetheless the factual content of that belief is a tautology and is therefore trivial.[32] He is thinking here of his account of the tenseless truth conditions of tensed sentences. Mellor maintains that the belief

> A. The experiences which I am now having possess the property of being present

is just true by definition on the New Tenseless Theory of Language. For the truth conditions of (A) are given by

> B. The experiences which I have at the time of the token of (A) possess the property of existing at the time of the token of (A).

But (B) is trivially true, says Mellor, a mere tautology. Therefore, although (A) is true, its factual content, as disclosed by (B), does not imply the objective reality of presentness.

This response by Mellor is multiply flawed. First, Mellor's tautology is self-constructed, for he stipulates that it is the experiences which I am now having which are judged to be present. But there is no reason to describe one's experiences as those which one is now having. The beliefs in question are not like (A); rather they are like

> A'. My experience of seeing the supernova is present,

which is not tautologous.

Second, even (A) can be read in a way that is not tautologous. Let the phrase "the experience which I am now having" pick out a specific, unique experience such as observing the supernova. In that case, the ascription of presentness to that particular experience out of all the experiences one ever has is not trivial or true by definition.

Third, even if (A) is trivial, that does not imply that the presentness of experience is trivial. It may be trivial to assert that "My present experiences

---

[32] Ibid., 54; see also his revised account in D. H. Mellor, "MacBeath's Soluble Aspirin," *Ratio* 25 (1983): 92.

are present" or that "My present experiences are experiences." But that does nothing to explain away the fact that one does have present experiences or to defeat the belief in the presentness of one's experiences.

Fourth, stating tenseless truth conditions for one's belief in the presentness of one's experience does not constitute even a *prima facie* defeater of that belief. Such truth conditions are just irrelevant to the proper basicality of that belief. For the object of one's belief is not the fact which is stated as the tenseless truth conditions of what one believes. In order for that to be the case the statement of the truth conditions would have to have the same *meaning* as the statement of the tensed belief, which is to lapse back into the Old Tenseless Theory of Language. Since they are not synonymous, the triviality of the statement of the truth conditions does not imply the triviality of the tensed belief. Nor is there any reason to think that the factual content of the tensed belief is given exhaustively in the tenseless truth conditions.

Finally, fifth, we have already seen (in the previous section) the shortcomings of Mellor's New Tenseless Theory of Language. Since his account of the presentness of experience is founded on the New Tenseless Theory, the demise of that theory also pulls under Mellor's account of the presentness of experience.

It therefore seems to me that Mellor has not provided a successful defeater of our belief that our experiences are present. Not only does such a belief seem to be properly basic, but it even seems to be indefeasibly true.

Our Differential Attitudes toward Past and Future

A second way in which we experience the reality of tense is exhibited by our attitudes toward the past and the future. We recall past events with nostalgia or regret, depending on whether they are remembered as pleasant or unpleasant, whereas we look forward to future events with either dread or anticipation. The beliefs that these attitudes express are tensed beliefs. As the late Oxford tense logician A. N. Prior once remarked, when we say, "Thank goodness that's over!" we certainly do not mean "Thank goodness the date of that thing's conclusion is June 15, 1954!" or "Thank goodness that thing's conclusion is simultaneous with this utterance!"—for why should anyone thank goodness for that?[33] Prior's point is that such attitudes cannot concern tenseless facts but are about tensed facts. The further point is that it is entirely rational to have such attitudes. Therefore, the tensed beliefs evinced by these

---

[33] A. N. Prior, "Thank Goodness That's Over," *Philosophy* 34 (1959): 17.

attitudes must be rational as well. If it is rational for me to be relieved that my visit to the dentist is past, then my belief that my visit is past is also rational.

On the static theory of time, feelings of relief and anticipation must be ultimately regarded as irrational, since events really are not past or future. Yet one can safely say that no static time theorist has ever succeeded in divesting himself of such feelings. Indeed, anyone who did succeed in ridding himself of such feelings and the tensed beliefs they express would cease to be human.

In response to Prior, Mellor and MacBeath concede that such attitudes do express tensed beliefs; but they again have recourse to the New Tenseless Theory of Language to strip those beliefs of any tensed factual content. Mellor writes, "Thus I thank goodness my headache is over not because it is over but because I believe it to be over: and the content of this belief is fixed by its token-reflexive truth conditions. . . ."[34] Thus, my truly believing that my headache is over does not imply that my headache is objectively past.

Now certainly Mellor and MacBeath are correct that what my attitudes immediately express are tensed *beliefs,* not tensed *facts.* For the dreaded event may be avoided and so never come to pass at all. Or I might be relieved about something due to a false report. All this proves is that one's tensed beliefs are defeasible. But many times my tensed beliefs are correct. Indeed, sometimes they are indefeasibly correct, as when I believe that the pain I felt is over. And, as we have seen, contrary to Mellor and MacBeath, neither the factual content nor the truth of my tensed beliefs is fixed by the tenseless facts which are stated as their truth conditions according to the New Tenseless Theory. In other words, the question comes down once more to the presentness of experience. When I feel relief, what I am relieved about can be analyzed as a complex fact involving the beliefs that (i) my experience is present and (ii) some event is earlier than the present. I can be mistaken about (ii), but I cannot be mistaken about (i), and thus the objectivity of tense remains.

There is a further feature of our attitudes toward the past and future that deserves to be highlighted, namely, the *difference* in how we regard an event depending on its pastness or futurity. An unpleasant experience which lies in the future occasions feelings of dread; but that very same experience, once past, evokes feelings of relief. On a dynamic theory of time these different attitudes are grounded in the reality of temporal becoming. A future event has yet to exist and will be present; but a past event no longer exists and was present. Therefore, it is rational to have different feelings about these events.

---

[34] Mellor, "MacBeath's Aspirin," 91; see also Murray MacBeath, "Mellor's Emeritus Headache," *Ratio* 25 (1983): 86-87.

But on a static theory of time, this difference in attitude toward the past and future is groundless and, hence, irrational. As philosopher of time George Schlesinger points out, on the static theory of time there is no more difference between an event's being located one hour later versus one hour earlier than now, than there is in an event's being located one mile to the right versus one mile to the left of here, for neither "now" nor "here" is objective.[35] Whether past or future, both events are equally real; there is no temporal becoming; nor am I moving toward one event and away from the other; and the distinction between past and future is purely subjective. Therefore, it just makes no sense to look upon these events differently. And yet, as Schlesinger observes, such a differential concern is a universal human experience.

Think, for example, of the difference in one's attitude toward one's birth and one's death. On the static theory of time the period of personal non-existence which lies after one's death is of no more significance than the period of personal non-existence which lies before one's birth. And yet we celebrate birthdays whereas we typically dread dying, a dread that runs so deep that one's death, wholly in contrast to one's birth, seems to put a question mark behind the value of life itself. Many existentialist philosophers have said that life becomes absurd in light of "my death"; but no one has said this with regard to "my birth."

Defenders of static time have naturally been reluctant to dismiss as irrational our differing attitudes toward past and future events and so have instead tried to find some basis for this difference in the static theory. For example, Nathan Oaklander, an ardent defender of static time, insists that such a difference is rational because on the static theory time is asymmetric, that is to say, it has a direction as determined by the ordering of events according to the relations *earlier than/later than*.[36] Oaklander thinks that it makes all the difference in the world whether an event is later than one's location in time or earlier than one's location in time.

But it is evident, I think, that on a static theory of time the mere asymmetry of time is not an adequate substitute for temporal becoming. Stripped of all tense, the relations of *earlier than/later than* with respect to some event no more justify differing attitudes on my part than would the relations *to the right of/to the left of*. Indeed, on the static theory of time, there are really two directions to time: one the "earlier than" direction and the other the "later than" direction. In the absence of temporal becoming it is wholly arbitrary

---

[35] George Schlesinger, *Aspects of Time* (Indianapolis: Hackett, 1980), 35.
[36] L. Nathan Oaklander, *Temporal Relations and Temporal Becoming* (Lanham, Md.: University Press of America, 1984), 146.

how these directions are laid on the series of events. The two arrows of time could be turned 180 degrees without any inconsistency with the facts. Although some scientists try to appeal to the laws of thermodynamics or other physical processes to establish "the" single arrow of time, philosopher of science Lawrence Sklar points out that all such attempts *presuppose* a prior choice of direction—for example, that the direction of entropy increase is the "later than" direction.[37] In the absence of temporal becoming, such a choice is wholly arbitrary. We could have called the direction of entropy increase "earlier than" if we had wanted to. Thus, "earlier" and "later" simply do not have the significance on a static theory of time that they do on a dynamic theory.

Our differing attitudes toward past and future events serve to underline how deeply ingrained and how strongly held our tensed beliefs are. If the static theory of time is correct, feelings of relief, nostalgia, dread, and anticipation are all irrational. Since such feelings are ineradicable, the static theory would condemn us all to irrationality. In the absence of any defeater for our belief in the objective distinction between past, present, and future, such a belief remains properly basic and the feelings they evoke entirely appropriate.

The Experience of Temporal Becoming

A third and final feature of our temporal experience which deserves mention is our experience of temporal becoming. The fact of temporal becoming is as obvious as the existence of the external world. For we experience that world as a continuous flux. Thus our experience of the external world is an experience of temporal becoming. But the reality of temporal becoming is even more evident to us than the reality of the external world. For in the inner life of the mind we experience a continual change of the contents of consciousness, and this stream of consciousness, even in the absence of any apprehension of the external world, makes evident to us the reality of temporal becoming. In the flux of experience there is constant and ineluctable becoming. It is therefore hard to imagine anything more obvious to us than the reality of temporal becoming.

The belief in temporal becoming comes to expression in certain experiences that are common to human beings. For example, who among us has not wished that it were some other time? A child anticipating Christmas morning might exclaim, "Oh, I wish it were Christmas!" Or someone going through hard times might think back on better times and say, "I wish it were 1968!" As Schlesinger points out, although there is no chance of such a wish's

---

[37] Lawrence Sklar, *Space, Time, and Space-Time* (Berkeley: University of California Press, 1976), section F of chapter 5.

being fulfilled, there is no lack of clarity as to what is being wished: "Anybody familiar with my plight would fully sympathize with me and unfailingly grasp what feature of the universe I should like to be different from what it is: instead of the NOW being at t_1, I should like it to be at t_0."[38] In such experiences we wish that some other moment in time were present rather than the one which is. We thereby presuppose the reality of temporal becoming, since our wish expresses our belief in a changing and objective present.

Since on the static theory of time there is no objective present, any informed person (including the static time theorist) who expresses such wishes is irrational. The best that the defender of tenseless time can do to make sense of such experiences is to offer tenseless substitutes for these wishes, such as "I wish Christmas were celebrated on December 1 instead of December 25," or "I wish the events of the world were reconstituted so that the world would appear to be the way it looked in 1968." But these things are obviously not at all what we are wishing for! The tenseless time theorist thus seems obliged to say that our real wishes, which are probably universal experiences of mankind, are just irrational.

Not so, retorts Oaklander. He acknowledges that "such wishes are meaningful" and admits that if the defender of the static view of time is committed to regarding such wishes as devoid of meaning, then there is "something wrong" with the static view of time.[39] Oaklander grants that the tenseless view of time cannot account for the meaningfulness of a wish that the "now" be located elsewhere than it is. But, he insists, this is not the meaning of my wish when I say, "I wish it were 1968." So what is the meaning of my wish? Oaklander answers, "I would be wishing that I could be perceiving and not merely remembering these things I perceived ten years ago. . . . That is, I wish that I was now perceiving events that are quite other than those I am in fact now perceiving."[40]

But it seems quite obvious that Oaklander's reconstruction of my wish is not at all what I am hoping for. For my wish has nothing to do with my perceptions—if I wanted to have different perceptions, I could go to a hypnotist! I want it to *be* 1968, not just to *appear* to be 1968. Schlesinger seems to have properly understood my wish as the desire that some other time should be present. And Oaklander admits that he cannot accommodate such a wish.

But Oaklander is not through yet. For he distinguishes the meaning of

---

[38] Schlesinger, *Aspects of Time*, 39.
[39] Oaklander, *Temporal Relations*, 159.
[40] Ibid., 160.

my wishes from the conditions under which my wish would be fulfilled. He says, "my wish that it was t_0 would be satisfied if and only if the thought that is the wish corresponds to the fact that *that thought* is simultaneous with . . . the time t_0."[41] The reader will by now recognize this as yet another appearance of the New Tenseless Theory of Language: It would be 1968 if and only if a token of that wish *occurs* in 1968. Not only do all our former objections serve to drive this poor player from the stage, but his appearance at this point makes no sense anyway. For my wish does not *mean* what is stated as its tenseless truth conditions, and therefore it is irrational on the static theory for me to wish for what I wish for rather than to wish for the tenseless fact expressed in its truth conditions.

Moreover, Oaklander's recourse to the New Tenseless Theory actually serves to expose a further weakness in that theory, namely, that theory cannot in fact tell us under what conditions my wish would be true. The conditions cannot be that a token of my wish *occurs* in 1968, for if it were 1968 then I would not have any such wish and so there would be no token! This is even more obvious if someone were to entertain the wish, "I wish it were the Cretaceous Period." For if it were the Cretaceous Period, there would not be anyone around to do the wishing! Thus, Oaklander is stuck with a meaningful wish, which can be rationally entertained, for which he can provide no conditions under which it would be fulfilled. By contrast the advocate of a dynamic theory of time can offer the straightforward account that my wish would be fulfilled if and only if it were now 1968, that is to say, if it were true that "1968 is present."

Another universal human experience that presupposes the reality of temporal becoming is the experience of waiting. When we wait for something to happen, we are experiencing the lapse of time in anticipation of some event. We do not merely experience the tenseless length of the temporal interval between our location and the location of the later event. Nor is it enough simply to occupy all the temporal locations between one's location and the location of the later event (even an inanimate object does that!) Rather there must be the experience of the passage of time. In the experience of waiting we apprehend temporal becoming, as things come to be and pass away until the anticipated event occurs. If the static view of time were correct, it would be irrational to wait for anything, since there is no temporal becoming. But such an experience is unavoidable.

About all the defender of static time could do at this point is to offer

---

[41] Ibid., 161.

tenseless, token-reflexive truth conditions for the beliefs presupposed by the experience of waiting—a response which we have seen to be irrelevant and unavailing.

In summary, then, a phenomenological analysis of our temporal experience reveals that we experience events as happening presently, that we have peculiar attitudes toward an event depending on whether it is past or future, and that we experience temporal becoming. No doubt there are many other examples of the way in which our belief in the objective reality of tense is manifested in our experience. But these examples serve well to show how basic, deeply ingrained, strongly held, and universal is our belief in the reality of tense and temporal becoming. On a static theory of time we are all of us hopelessly mired in irrationality, prisoners to an illusion from which we are powerless to free ourselves. By contrast, if a dynamic theory of time is correct, our experiences and beliefs are entirely rational and appropriate. Thus, insofar as we think that such experiences are justified, we should embrace a dynamic theory of time.

It follows from the above argument that we are *prima facie* justified in holding our belief in the objective reality of the distinction between past, present, and future. Far from being controversial, such a conclusion could be accepted even by a proponent of a tenseless view of time. For we have yet to consider arguments *against* a dynamic theory of time, which may serve to remove the *prima facie* justification accorded to our tensed beliefs by experience and so defeat the current argument. Mellor, for example, despite all his objections, frankly admits, "Tense is so striking an aspect of reality that only the most compelling argument justifies denying it: namely, that the tensed view of time is self-contradictory and so cannot be true."[42] That is why McTaggart's Paradox—to be considered in the next section—constitutes, in Mellor's own words, "the lynchpin of my book."[43] If McTaggart's Paradox fails to defeat belief in the objective reality of tense, then Mellor admits that we are justified in holding our tensed beliefs.

We shall consider McTaggart's Paradox below, but before we conclude this section it is worth pausing a moment to reflect on the strength of the argument thus far and on what it would take to defeat it. Defeaters are beliefs which are incompatible with some belief we hold and which have more warrant than our current belief. If we are to be rational in the face of an alleged defeater, we must either abandon our original belief or else defeat the defeater

---

[42] Mellor, *Real Time*, 4-5.
[43] Ibid., 3.

itself. One way to defeat an alleged defeater is to find some third belief which is compatible with our original belief but incompatible with the alleged defeater, and which has even more warrant than the alleged defeater. This third belief would be an *extrinsic* defeater-defeater. But there is also such a thing as an *intrinsic* defeater-defeater. In this case the original belief itself is seen to have more warrant than the defeaters brought against it, and so it simply overwhelms its alleged defeaters. For example, suppose you were accused of a crime which you knew you did not commit, even though all the evidence stood against you. Would you be rationally obliged in such a case to abandon belief in your innocence and go along with the evidence and believe that you are guilty? Of course not! Your belief that you did not commit the crime has far more warrant for you than the belief that you are guilty, despite the evidence supporting that accusation. Your belief intrinsically defeats its alleged defeater.

Now it deserves to be asked at this point whether our belief in the reality of tense and temporal becoming is not so powerfully warranted that it becomes an intrinsic defeater of the defeaters brought against it. On the basis of our phenomenology of temporal consciousness, it is hard to see how any belief could be more warranted for us than, say, our belief in the presentness of experience. What argument for the unreality of tense could possibly be based on premises more evident than that basic belief? McTaggart's Paradox? Hardly! In the face of our basic belief in the reality of tense and temporal becoming, that paradox—even in the absence of a resolution—takes on the air of Zeno's arguments for the impossibility of motion: an engaging and recalcitrant brain-teaser whose conclusion no one really takes seriously. I suspect that we shall find ourselves far more certain of the reality of tense than of the cogency of McTaggart's argument.

We thus have two very powerful arguments, the linguistic argument and the phenomenological argument, in favor of the tensed theory of time. It now remains to be seen what arguments can be brought against that theory.

II. Arguments against a Dynamic Conception

1. *McTaggart's Paradox*

EXPOSITION

In 1908 the Cambridge idealist John Ellis McTaggart published a remarkable article in the journal *Mind* entitled "The Unreality of Time."[44] McTaggart

---

[44] J. Ellis McTaggart, "The Unreality of Time," *Mind* 17 (1908): 457-474. The argument is defended against objections in McTaggart's *The Nature of Existence*, 2 vols., ed. C. D. Broad (Cambridge: Cambridge University Press, 1927; rep. ed.: 1968), chapter 33.

was not kidding: He firmly believed that he had come up with an argument which proves that time does not exist, and this argument was the principal legacy he bequeathed to twentieth-century philosophy. The philosopher of time Richard Gale has observed, "If one looks carefully into the multitudinous writings on time by analysts, one can detect a common underlying problem, that being that almost all of them were attempting to answer McTaggart's paradox."[45]

What is McTaggart's Paradox? The argument consists of two parts. In the first part McTaggart argues that time is essentially tensed. In the second part he argues that tensed time is self-contradictory. It therefore follows that time is unreal.

The first to distinguish clearly between tensed and tenseless views of time, McTaggart thus has something for everyone. Tensers love the first part of his argument, that time is essentially tensed, but they disagree with the second part, that tensed time is self-contradictory. De-tensers love the second part of his argument because it shows that the tensed view of time cannot be true, but they reject the first part because they think that time is in fact tenseless. Virtually no one agrees with McTaggart himself that time is unreal; rather the question has become the *nature* of time: Is it tensed or tenseless?

Since our concern is with arguments against a tensed or dynamic theory of time, we shall focus on the second half of McTaggart's proof. His argument here is apt to appear bewildering unless we first understand its metaphysical presuppositions. The key to understanding the contradiction McTaggart sees in a dynamic view of time is his presupposition that past, present, and future events are all equally real or existent and that temporal becoming consists in the movement of the present along this series.[46] McTaggart thinks of the series of temporal events as stretched out like a string of light bulbs which are each momentarily illuminated in succession, so that the light is seen to move across the series of bulbs. In the same way presentness moves across the series of events. Since all events are equally existent, the only respect in which they change is the change in tense that they undergo. First they are future, then they are present, then they are past. In every other respect they just *are*. Obviously, then, for McTaggart, *becoming present* does not imply *becoming existent*.

McTaggart observes that pastness, presentness, and futurity are mutually incompatible: No event can have all three. But given McTaggart's tenselessly

---
[45] Gale, *Language of Time*, 6.
[46] See, for example, his statements in McTaggart, "Unreality of Time," 463; idem, *Nature of Existence*, 2:11.

existing series of temporal events, every event does have all three! Take an event tenselessly located at t_1. At t_1 that event is obviously present. But because all events are equally real, that same event also has pastness and futurity because at t_2 it is past and at t_0 it is future. The moment t_1 is not any more real or privileged than t_0 or t_2, and so the event in question must be characterized by the tenses it has at all these times, which is impossible. We can visualize the problem by imagining the people existing at each of these three moments. For the people at t_1, t_1 is present. Since neither t_1 nor these people pass away, it is still the case when it is t_2 that, for the people at t_1, the moment t_1 is present. But for the people at t_2, the moment t_1 is past. The moment t_1 never sheds presentness and takes on pastness—just ask the people at t_1! But t_1 never exchanges its pastness for any other tense either, as the people at t_2 will tell you. Thus, t_1 is changelessly both present and past, which is impossible. If someone should say, "But t_1 is present relative to t_1 and past relative to t_2, which is not contradictory," the advocate of tenseless time will say that such relational properties reduce to the tenseless relations *is simultaneous with* and *is earlier than*, which vindicates the tenseless theory.

Perhaps another way of getting at the difficulty McTaggart sees is to ask *when* t_1 has presentness. The answer can only be: at t_1. But t_1 always has presentness at t_1—that is tenselessly true! Therefore, t_1 never changes its tense if we say that t_1 has presentness at t_1 (and pastness at t_2, and futurity at t_0, for these assertions are all tenselessly true). But if events never change their tense, then either time does not exist (McTaggart's conclusion) or the static theory of time is correct (the de-tenser's conclusion).

McTaggart observes that someone will respond that t_1 does not merely have presentness at t_1 but just has presentness, period. When t_1 is absolutely present, then it is not past or future as well, so that no contradiction arises. But McTaggart rejects this response because it leads to a vicious infinite regress of hyper-times, as illustrated in Fig. 4.1. In hyper-time, presentness moves along the series of moments of time. In this way one can make sense of, say, t_1's not only being present at t_1 but also being absolutely present. For this absolute present is the present of hyper-time, in which all the moments of time are embedded.

Now the postulation of an embedding hyper-time might in itself seem so metaphysically extravagant that the reality of tense should be rejected. But McTaggart's objection is even more fundamental: The postulation of a hyper-time solves nothing. For since hyper-time is also tensed, the same contradiction arises again on the hyper-level. The moments of hyper-time must all be equally real and therefore must each be past, present, and future, which is

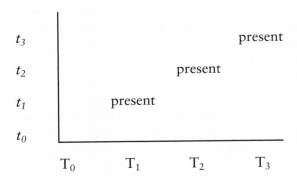

Fig. 4.1: At successive moments of hyper-time T, successive moments of time t become present. Thus, for example, t_2 becomes present at T_2.

impossible. The only way to escape this contradiction is to posit a third-level hyper-hyper-time in which the moments of hyper-time become successively present (Fig. 4.2).

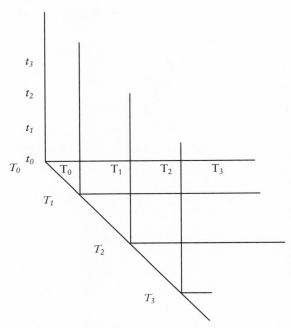

Fig. 4.2: Hyper-time T must be embedded in a hyper-hyper-time $\mathcal{T}$ in which moments of T become successively present.

But obviously the same problem will recur on this third level, and so one must posit yet another level to resolve it, on to infinity. This sort of infinite regress is what philosophers call a "vicious" infinite regress, because at every level the problem remains unsolved. Thus, it simply does no good to try to elude McTaggart's Paradox by claiming that moments of time are successively future, then present, then past.

We can formulate McTaggart's Paradox as follows:

1. If a tensed view of time is correct, events are past, present, and future.

2. Events cannot be past, present, and future unless either

   i. events are past, present, and future only in relation to other events,

or

   ii. events are past, present, and future in hyper-time.

3. If a tensed view of time is correct, then events cannot be past, present, and future only in relation to other events.

4. If a tensed view of time is correct, then events cannot be past, present, and future in hyper-time.

5. Therefore, if a tensed view of time is correct, events cannot be past, present, and future.

6. Therefore, if a tensed view of time is correct, events are past, present, and future, and events cannot be past, present, and future.

7. Therefore, a tensed view of time is not correct.

Critique

Some philosophers have sought to avoid McTaggart's Paradox by denying premise (1). They claim that it is question-begging to assert that events *are* (tenselessly) past, present, and future. Rather we must tense the verb and say that events are present, were future, and will be past. Then there is no contradiction. Now while I think these philosophers are on to something important, such a response is not relevant to the argument as I have formulated it. For the contradiction deduced in premise (6) is not that events are past, present, and future, but rather that they both are and are not past, present, and future. The tenseless language employed in (1) is just a harmless *façon de parler*. It just expresses the truth that on a tensed view of time events change with respect to their tense. The real issue is how they can do this, which is addressed in premise (2). Now McTaggart seems to me to be quite correct that the alternatives (i) and (ii) will not work for the tensed theory of time, and therefore premises (3) and (4) are true. If McTaggart has erred, then, it is by omitting some alternative from premise (2) and leaving us with a false dilemma.

It seems to me that this is exactly what has happened. McTaggart's whole argument is based on a misguided attempt to marry a dynamic theory of temporal becoming to a static series of events. It is no wonder that the dynamic-static theory of time he winds up with proves to be self-contradictory! Sharp-sighted critics of McTaggart such as C. D. Broad and A. N. Prior have insisted almost from the beginning that a dynamic or tensed theory of time implies a commitment to *presentism,* the doctrine that the only temporal entities that exist are present entities.[47] According to presentism, past and future entities do not exist. Thus, there really are no past or future events, except in the sense that there have been certain events and there will be certain others; the only real events are present events. Thus, there can be no question of an event's swapping futurity for presentness or cashing in presentness for pastness. Temporal becoming is not the exchange of tense on the part of tenselessly existing events but the coming into and going out of existence of the entities themselves. Events no more change tenses than they exchange properties of non-existence and existence! McTaggart's Paradox thus arises, not from a dynamic theory of time, but from a misconceived union of the dynamic and static theories of time.

That presentism does avert McTaggart's Paradox is evident from recent philosophical discussions of intrinsic change.[48] The problem posed by intrinsic change is basically this: How can something possess different properties at different times and yet be the same, identical thing? Necessarily, if "two"

---

[47] C. D. Broad, *An Examination of McTaggart's Philosophy,* 2 vols. (Cambridge: Cambridge University Press, 1938; rep. ed.: New York: Octagon Books, 1976), 2:280-302; A. N. Prior, "The Notion of the Present," *Studium Generale* 23 (1970): 245-248.

[48] See David Lewis, *On the Plurality of Worlds* (Oxford: Basil Blackwell, 1986), 203-204; Trenton Merricks, "Endurance and Indiscernibility," *Journal of Philosophy* 91 (1994): 165-184.

things are identical, then they have all the same properties (after all, they are the same thing!). But then how can something which exists at t_1 be identical with something which existed at t_0 unless they have all the same properties? How is change over time possible, or, to put the question another way, how is identity over time possible?

Now McTaggart's Paradox is actually a peculiar instance of this problem of intrinsic change. For it asks how some event E can be the same event if at t_1 it has presentness whereas at t_2 it has pastness. Or to put it another way, if E as perceived at t_1 is the same, identical event which is remembered at t_2, then how can E possess different properties or tenses at t_1 and t_2? If one says that E's tense changed from t_1 to t_2, then it is not E that one remembers after all, but some different event. Thus, McTaggart's argument is that intrinsic changes in the tense of events is impossible.

The presentist eludes the problem of intrinsic change by holding that the only properties a thing possesses are those it presently possesses. Thus, if a thing was red at t_0 and is blue at t_1, it does not have (present tense) incompatible properties. For it only has the properties it presently has, including *blueness*. It did have *redness* once, but no longer. Thus, the thing which existed at t_0 has (present tense) all the same properties which it has (present tense) at t_1. There is thus no contradiction in intrinsic change.

Similarly, an event possesses only the tense it presently has, namely, presentness. No event ever possesses pastness or futurity, for non-present events do not exist. Thus, there can be no question of any event's possessing incompatible tense determinations. This is the germ of truth in some philosophers' denial of premise (1). Events only have presentness; but they can be said to be past or future in the sense that it is (presently) true that they were once present or will be present. Thus, McTaggart's Paradox is ineffectual against the presentist.

It is therefore instructive to observe that contemporary proponents of McTaggart's Paradox, despite their varying formulations of the problem, all presuppose a hybrid dynamic-static theory of time, just as McTaggart did. For example, Mellor asserts, "Futurity, temporal presence, and pastness are all supposed to be real non-relational properties which everything in time successively possesses, changing objectively as it exchanges each of these properties for the next."[49] Similarly, Mellor's former student and collaborator R. Le Poidevin insists that even those who deny that the future is real imply by that very denial that the past, at least, is real: "although the dinosaurs (for

---

[49] Mellor, *Real Time*, 4.

example) are extinct, they are still real to the extent that it is they and their properties which make statements about dinosaurs true. . . . To make sense of the past's being real and the future not, we have to talk of being real *simpliciter,* not once being real, or being about to be real."[50] Oaklander could not be clearer in his misconstruing the tensed theory of time as a union of the dynamic and static theories:

> On the *traditional* tensed . . . theory of time, the NOW is a particular or property that moves along an ordered, but as yet non-temporal, . . . series. The terms of the . . . series exist (tenselessly) in unchanging relations to each other, and these unchanging relations become temporal relations as the NOW moves across them. . . .[51]

What our discussion reveals, then, is that McTaggart's Paradox is not really an argument against a dynamic theory of time at all. Rather it is aimed (and effectively so, I think) at a teratological hybrid, the dynamic-static theory of time.

Thus, if we adopt presentism in line with a pure dynamic theory of time, we avoid McTaggart's Paradox. Significantly, Le Poidevin, at least, admits this: Presentism "represents the only means to block McTaggart's proof of the unreality of time consistently with the assumption of a non-relational past, present, and future."[52] Presentism enables us to revise premise (2) so as to admit a third alternative:

2'. Events cannot be past, present, and future unless either

    i. events are past, present, and future only in relation to other events,

or

    ii. events are past, present, and future in hyper-time,

or

    iii. only present events exist, and events are past or future only in the sense that they were or will be present.

---

[50] Robin Le Poidevin, "Lowe on McTaggart," *Mind* 102 (1993): 168.
[51] L. Nathan Oaklander, "Zeilicovici on Temporal Becoming," *Philosophia* 21 (1991): 329.
[52] Robin Le Poidevin, *Change, Cause, and Contradiction: A Defense of the Tenseless Theory of Time* (London: Macmillan, 1991), 36.

The proponent of McTaggart's Paradox, if his argument is to be successful, must now refute the presentist alternative (iii). So what is wrong with presentism?

Presentism is frequently rejected because it is thought to imply, in conjunction with the Special Theory of Relativity, a sort of solipsism (the view that I alone exist), which no sane person can believe.[53] This unwelcome consequence is due to the absence of absolute time and space within the context of STR, which makes it impossible to define any plausible co-existence relation between oneself and other things.[54] Although we shall have more to say about this subject later,[55] anyone who has followed our argument in chapter 2 will realize that this objection to presentism is easy to answer. The objection is predicated upon an Einsteinian interpretation of STR which the defender of divine temporality should reject on wholly independent grounds in favor of a Lorentzian interpretation. A Lorentzian understanding of relativity, it will be recalled,[56] preserves relations of absolute simultaneity and so confronts no challenge concerning co-existence relations among temporal beings. The presentist who accepts Lorentzian relativity is thus not threatened by the specter of solipsism.

Le Poidevin also charges that presentism implies "temporal solipsism," but by that epithet he signifies a number of technical philosophical doctrines which he finds objectionable.[57] While I agree with Le Poidevin that these doctrines are implausible, his case that they are implied by presentism is not very convincing. For example, he claims that if presentism is true, then we cannot make true existence statements about the past, such as "Some Frenchmen fell at Waterloo." This is because the logical form of such a statement is understood to be *there is some individual* x *such that* x *was a Frenchman and* x *fell at Waterloo*. Notice that according to the logical form of this statement there is some individual *x*. Le Poidevin takes this to show that logic commits us to the reality of all individuals, whether they are past, present, or future.

This argument, however, strikes me as quite contrived. The language of logic is an artificial, tenseless language which simply ignores tense distinctions in ordinary language. The presentist agrees that, in the tenseless universe of discourse of classical logic, we can truly say that there *are* (tenselessly) past and

---

[53] See, for example, D. H. Mellor, "Special Relativity and Present Truth," *Analysis* 34 (1973–1974): 75-76.
[54] Yuri Balashov, "Enduring and Perduring Objects in Minkowski Space-Time," *Philosophical Studies* 99 (2000): 129-166.
[55] See chapter 5, pages 169-170.
[56] See chapter 3, pages 54-57.
[57] Le Poidevin, *Change, Cause, and Contradiction*, chapter 3.

future individuals, that is to say, logic ranges over all individuals in the actual world, abstracting from whether they are past, present, or future. There is just no metaphysical significance in this artificial tenseless discourse. Now if we do want to invest logical form with metaphysical significance, then the presentist can propose either of two reforms: (i) we can take existence claims to be, not tenseless, but multiply tensed; for example, *there was, is, or will be some individual* x *such that* x *was a Frenchman and* x *fell at Waterloo;* or else (ii) we may supplement classical logic with so-called tense logic by prefixing existence statements with tense indicators; for example, *it was the case that there is some individual* x . . . or *it will be the case that there is some individual* x . . . By means of either of these alternatives the presentist can make it clear that existence statements about past and future individuals do not imply that those individuals are as real as present individuals. But these complications are doubtlessly unnecessary, since there is just no reason to regard the artificial language of classical logic as fraught with metaphysical significance.

Again, Le Poidevin objects that if presentism is true, then there can be no relations between things which do not exist at the same time, since at least one of them does not exist. But obviously such relations do exist; for example, the Battle of Hastings was earlier than the Battle of Waterloo, Kennedy was envied by Nixon, Aquinas was smarter than Attila, and so forth. There are a couple of problems with this objection. First, it just assumes that trans-temporal relations cannot exist. If there are such things as relations, then why do both members of a relation have to exist at the same time? Second, the objection proves too much. If it were correct, then there could be no relations between individuals in different logically possible worlds. There could be no relation of trans-world identity, for example. Thus, we could not say that in some possible world I weigh one pound more than I actually do, which implies the absurd conclusion that I have all my properties essentially. Any philosophical reconstruction of such trans-world relations aimed at replacing them with more acceptable notions can be paralleled by a similar reconstruction of trans-temporal relations. They stand or fall together. Finally, third, in many cases we probably could dispense with trans-temporal relations. For example, we could simply say that Nixon had the property of *envying Kennedy*. No further relation is required. (Indeed, we can imagine cases where someone might be envious of, say, a wholly fictitious person, in which case there is no relation to an individual at all.) Or again, we could say that Attila had a certain IQ and Aquinas had a certain IQ, and the one number is higher than the other. Even relations of *earlier/later than* can be plausibly analyzed in non-relational terms, as we shall see.[58] Thus, Le Poidevin's second objection is not compelling.

Le Poidevin presents further difficulties, but these seem no more persuasive than the above, so the reader may be left to pursue the more technical discussion of these issues should he desire to.[59]

So far we have not seen compelling reasons to reject the presentist perspective. Indeed, considerations arising from a discussion of McTaggart's Paradox suggest a positive reason to accept presentism. For there is a sort of modal McTaggart's Paradox, which is parallel to the temporal McTaggart's Paradox, the solution to which is analogous to presentism.[60] It goes like this: Something cannot be both actual and merely possible. But everything that exists is both of these: It is actual in the actual world and merely possible in some other possible world in which it does not exist. Someone will say that it is not inconsistent to be actual relative to the actual world and merely possible relative to some other world. But this leads to an infinite regress. For since the actual world is actual relative to itself and some merely possible world is actual relative to itself, we must postulate some hyper-actual world in which only one of these two worlds is actual. But then the same problem arises for the hyper-actual world, and so on to infinity.

The usual answer to this modal problem is to adopt the doctrine of *actualism,* the view that only the actual world is real. Possible worlds are merely abstract ways the world might conceivably have been. Thus, there really are no concrete, parallel worlds which have actuality as does the actual world. Now actualism is precisely parallel to presentism. As Le Poidevin admits, "The doctrine that only the actual world is real avoids the modal paradox just as the doctrine that only the present is real avoids McTaggart's paradox."[61] Since these two problems and their solutions are parallel, consistency demands that they must be accepted or rejected together. Either accept both actualism and presentism or else hold that just as all moments of time are equally real, so all possible worlds are equally real. The same sort of objections that Le Poidevin lodges against "temporal solipsism" can be analogously lodged against "modal solipsism"; and the same sort of moves that the actualist can make to escape these can also be made by the presentist.

---

[58] See chapter 5, pages 190-192.
[59] See my "McTaggart's Paradox and Temporal Solipsism," *Australasian Journal of Philosophy* 79 (2001): 32-44.
[60] See M. J. Cresswell, "Modality and Mellor's McTaggart," *Studia Logica* 49 (1990): 163-170.
[61] Le Poidevin, *Change, Cause, and Contradiction,* 35.

Thus, thinkers such as Le Poidevin who want to embrace actualism and yet reject presentism find themselves in real tension.

It only needs to be added that the tiny handful of philosophers who consistently believe in the concrete reality of all possible worlds and all moments of time find themselves burdened with a metaphysical worldview which the vast majority of philosophers find quite outlandish.[62]

Thus, the key to avoiding McTaggart's Paradox is presentism. McTaggart's central mistake, as the distinguished British philosopher Michael Dummett has pointed out, is that he assumed that there must exist a single, complete description of reality.[63] But if we take tense seriously, no such description can exist. There is rather a different description of reality that holds at each and every moment that is present. McTaggart wants to describe the world, as the medieval theologians put it, *sub specie aeternitatis* (from the standpoint of eternity), and yet include tense in that description. You cannot do both. A timeless view of reality excludes tense, which is why the reality of tense and temporal becoming implies that God, as One who is really related to and knows the world, is temporal. One can hardly, then, object to the reality of tense, as McTaggart does, on the grounds that a tensed world cannot be captured in a single, timeless description.

2. The Myth of Passage

EXPOSITION

The idea that time flows or passes is a common idea in Western philosophy, at least as old as the ancient Greek philosopher Heraclitus. Isaac Newton, it will be recalled, held that absolute time "flows equably without relation to anything external."[64] McTaggart took such language literally. He states, "The movement of time consists in the fact that later and later terms pass into the present, or—which is the same fact expressed another way—that presentness passes to later and later terms."[65]

The passage of time is an undisputed feature of psychological time. During bursts of activity time seems to pass quickly, and we are apt to exclaim, "How time flies!" By contrast, when we are languishing, time passes

---

[62] For a critique, see Peter van Inwagen, "Indexicality and Actuality," *Philosophical Review* 89 (1980): 415-417.

[63] Michael Dummett, "A Defense of McTaggart's Proof of the Unreality of Time," *Philosophical Review* 69 (1960): 503; idem, "The Reality of the Past," *Proceedings of the Aristotelian Society* 69 (1968–1969): 252-253.

[64] Isaac Newton, *Sir Isaac Newton's "Mathematical Principles of Natural Philosophy" and his "System of the World,"* trans. Andrew Motte, rev. with an appendix by Florian Cajori, 2 vols. (Los Angeles: University of California Press, 1966), 1:6.

[65] McTaggart, *Nature of Existence*, 2:10.

excruciatingly slowly, and we complain, "Time keeps dragging by." The question is whether this feature of psychological time is also a feature of time itself.

A number of proponents of the static view of time have charged that the idea of a literal passage of time is nonsense and that therefore a dynamic theory of time cannot be correct.[66] For since the dynamic view of time is committed to the objective reality of temporal becoming, it implies the reality of the passage of time. Since time's passage is purely psychological, the dynamic theory does not give us the truth about time.

The passage of time must be a myth, it is argued, because otherwise unanswerable questions arise. For example, how fast does time flow? In cases of literal motion, we measure the distance traversed per unit of time; for example, sixty miles per hour. But in the case of time's passage, we are measuring the amount of time crossed per—what? What sense is there in talking about how fast a minute passes? A minute passes in a minute—a mere tautology! Thus, no non-trivial content can be given to the claim that time passes. Second, an event occurring at a single instant would have a sort of history if time passes: First it would be future, then it would be present, then it would be past. But since it only exists at an instant, it cannot have a history in ordinary time. Therefore, the passage of time must occur relative to a hyper-time. In so many units of hyper-time, presentness moves across so many units of ordinary time. But then we have to ask about the flow of hyper-time, and off we go on a vicious infinite regress. A literal flow of time is therefore incoherent.

We can formulate the objection to time's passage as follows:

1. If time is dynamic, time's passage is a mind-independent reality.

2. Time's passage cannot be a mind-independent reality.

3. Therefore, time is not dynamic.

CRITIQUE

It is a curiosity of the philosophical discussion of this issue that the truth of premise (2), so loudly trumpeted by certain static time theorists, is accepted by the wide majority of defenders of dynamic time. Nor does this represent a concession on their part to the proponents of static time. Rather the objections to a literal flow of time were borrowed by critics such as D. C. Williams

---

[66] Donald C. Williams, "The Myth of Passage," *Journal of Philosophy* 48 (1951): 457-472; J. J. C. Smart, "The River of Time," *Mind* 58 (1949): 483-494; idem, "The Temporal Asymmetry of the World," *Analysis* 14 (1953–1954): 79-83.

and J. J. C. Smart from C. D. Broad's critique of McTaggart. Broad, a philosophical convert to a pure tensed theory of time, saw clearly the problems inherent in McTaggart's hybrid dynamic-static theory.[67] Temporal becoming, Broad insisted, should not be thought of as the literal motion of presentness along a series of tenselessly existing events. Otherwise one lands in precisely the conundrums explained above. Temporal becoming, on Broad's view, is not a qualitative change in an event. Becoming real is not like, say, becoming fat, for in temporal becoming there is no enduring subject which moves from the future into the present or from non-existence to existence. Rather temporal becoming is absolute becoming—not becoming this or that, but simply coming to be. An event is simply something's coming into existence.

Broad's presentism thus led him to deny premise (1), that a dynamic view of time implies a literal passage of time. It is ironic (and perhaps indicative of the sloppiness of their argumentation) that static time theorists misappropriated Broad's objections to time's passage in order to argue against the dynamic theory itself. Most dynamic theorists would agree with A. N. Prior when he said that the flow or passage of time "is just a metaphor"—albeit an important one.[68] According to Prior the flow of time is metaphorical because what it refers to is neither a genuine motion nor a genuine change; but the force of the metaphor can be explained by the objectivity of tensed facts. That is the reality behind the metaphor.

More recently, Smart has acknowledged that dynamic time theorists take the passage of time to be a metaphor for objective temporal becoming.[69] But he questions whether objective temporal becoming makes any more sense than the idea of time's passage. Normally, we speak of something's becoming this or that; but temporal becoming is conceived to be absolute. "In the pure becoming of an event," demands Smart, "what does the event become?"[70]

Smart's question is strangely misconceived, however, for he himself has repeatedly emphasized Broad's point that it is *things*, not *events*, that come to be; an event is just the coming to be of some thing or things.[71] If a thing can be said to become anything, it becomes actual or real. But this is not the acquisition of a new property in place of a (pseudo-) property of non-actuality or unreality. It just is the existing of the thing with all its properties. Smart does

---

[67] Broad, *Examination of McTaggart's Philosophy*, 2:277-280.
[68] Arthur N. Prior, "Changes in Events and Changes in Things," in *Papers on Time and Tense* (Oxford: Clarendon, 1968), 1.
[69] J. J. C. Smart, "Time and Becoming," in *Time and Cause*, ed. Peter van Inwagen, Philosophical Studies Series in Philosophy (Dordrecht: D. Reidel, 1980), 4.
[70] Ibid., 5.
[71] See Smart, "River of Time," 486; idem, "Temporal Asymmetry," 81; *Encyclopedia of Philosophy* (1967), s.v. "Time," by J. J. C. Smart.

proceed to acknowledge that it makes sense to say that in temporal becoming an event becomes present. But he finds this explanation unhelpful, since every event becomes present at some time or other. Smart has here lapsed back into thinking of all events in the temporal series as equally real and so as equally present at their respective times. But on the presentist view, presentness is had absolutely, not just relative to a time, and thus the only events that truly have presentness are the events currently happening. Smart's objections may have purchase against a McTaggartian, hybrid view of time, but they are irrelevant to a pure dynamic theory of temporal becoming.

A much more serious difficulty confronts the defender of objective temporal becoming, however, a conundrum about time that is at least as old as Aristotle.[72] This is the problem of the extent of the present. If only the present exists and the past and future are unreal, then reality seems to be reduced to a literal instant. An instant has by definition zero duration. But if things exist for literally zero amount of time, how is this different from not existing at all? The claim that only present things exist thus seems self-destructive. This problem so impressed Broad that late in life he himself actually embraced the existence of a hyper-time in which events that are instantaneous in ordinary time endure.[73]

Could the present be a mere instant? A great many philosophers see no problem in this idea. The present would be like an instantaneous slice of space-time. An instant would be what is called a degenerate interval, that is to say, an interval of zero duration. An instantaneous slice of space-time would be, for example, everything existing at precisely the instant we mark at 3:00 P.M. EST. Such an instantaneous state of physical reality would be described by all the statements true at that instant.

While an instantaneous state seems to make sense, however, it is not clear how such a conception of reality is to be united with temporal becoming. Put as simply as possible, the problem is that since instants have no immediate successors (between any two instants there is always an infinity of intermediate instants), it is difficult to see how time can elapse instant by instant, one at a time, consecutively. Moreover, how could any non-zero interval of time ever elapse, since the addition of durationless instants can never add up to a non-zero interval? These difficulties, which deeply disturbed a great mind like Alfred North Whitehead, are resolved by the eminent philosopher of space

---

[72] See Aristotle, *Physics* 4. 10. 217b33-218a9. For a wonderful discussion of the early history of this conundrum see Richard Sorabji, *Time, Creation, and the Continuum* (Ithaca, N.Y.: Cornell University Press, 1983), 7-63. Augustine in particular agonized over this problem in his *Confessions* 9.15-28.

[73] C. D. Broad, "A Reply to My Critics," in *The Philosophy of C. D. Broad,* ed. P. A. Schilpp, Library of Living Philosophers (New York: Tudor, 1959), 769-772.

and time Adolf Grünbaum only at the expense of denying the reality of temporal becoming and embracing a static theory of time.[74]

Whitehead preferred to deny that temporal becoming is a continuous process involving instants, advocating instead the existence of minimum, discrete "atoms" of time, often called "chronons." On an atomistic view, although time is infinitely divisible *in thought,* there are *in reality* indivisible, finite intervals of time which compose time. Chronons may or may not be conceived to have precise boundary points; rather than thinking of them like marbles in a line, we should perhaps think of them as blurry, shading into one another. On the atomist view, only the present chronon exists, being wholly present, and temporal becoming proceeds one chronon at a time.

One disturbing feature of such an understanding of temporal becoming is that becoming is "jerky" rather than smooth. Reality unfolds like the successive frames of a movie film projected on a screen—the frames pass too quickly for the discontinuities to be noticed, but there are "leaps" between them nonetheless. Not that there is anything that happens in between chronons which we miss out on, for there is no in-between. On such a view, change is discontinuous.

This can lead to some bizarre results. Consider, for example, the Stadium Paradox of the ancient Greek philosopher Zeno. He invites us to imagine two rows of spatial atoms moving in opposite directions along a row of atoms at rest at the rate of one atom per chronon (Fig. 4.3).

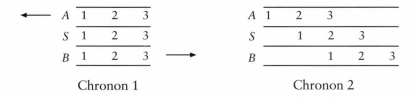

Fig 4.3: Zeno's Stadium Paradox

At chronon 1, A_1 and B_1 are aligned, then at chronon 2, A_3 and B_1 are aligned. But this seems crazy because it implies that there never was any event of the alignment of A_2 with B_1. But in order to move from being aligned with A_1 to

---

[74] See Alfred North Whitehead, *Process and Reality,* corr. ed., ed. David Ray Griffin and Donald W. Sherburne (New York: Free Press, 1978), 68; idem *Science and the Modern World* (New York: Macmillan, 1925), 125-127; Adolf Grünbaum, "Relativity and the Atomicity of Becoming," *Review of Metaphysics* 4 (1950–1951): 143-186; idem, "A Consistent Conception of the Extended Linear Continuum as an Aggregate of Unextended Elements," *Philosophy of Science* 19 (1952): 288-306.

being aligned with A_3, B_1 must have passed A_2 at some time. If we say that such an alignment did occur, it must have occurred in between chronons, which is impossible. Therefore, we have no choice but to say that reality jumped discontinuously from one state to another. If we can find some way of avoiding this sort of weirdness, it would surely be preferable to do so!

Let us therefore explore a different understanding of the extent of the present. We might maintain that "the present" is not a metrical concept and therefore does not refer to any specific measured interval. This alternative seems to have been implied by the French philosopher Henri Bergson.[75] In Bergson's view real duration is not composed of instants or time atoms but is prior in reality to our mathematization of it. All metrical concepts of time are secondary constructions. British philosopher Rom Harre thus contends that the question about the extent of the present

> only makes sense if we have already accepted a certain mathematical model for the formal representation of temporal discourse, and then have taken that model to be descriptive in all its aspects of some basic temporal reality.
>
> The model is a continuous, linear point manifold on to which the expressions that occur in temporal talk are to be mapped. . . . For example, the expressions "past," "present," and "future" are translated into mathematical features of the manifold.[76]

Reject that model, as Bergson does, and the question "How many points of the manifold does the 'now' enclose?" is no longer a question about reality.

On this view, to ask, "What is the extent of the present?" is a malformed question. In order for the question to be meaningful, one must stipulate what it is we are talking about: the present vibration of an atomic clock, the present session of Congress, the present war, or what have you? There is no such metric interval as "the present," period; we must speak of "the present ____," where the blank is filled by a reference to some event or thing. If we choose to speak of time itself, then our question becomes trivial: "How long is the present minute?" "One minute!"

If an atomist were to demand what the minimum temporal duration is, we could coherently reply that there is no minimum duration. Time should not be thought of as *composed* of an infinite number of instants; but any

---

[75] Henri Bergson, *Duration and Simultaneity*, trans. Leon Jacobson, with an introduction by Herbert Dingle, Library of Liberal Arts (Indianapolis: Bobbs-Merrill, 1965), chapter 3. See also Andros Loizou, *The Reality of Time* (Brookfield, Vt.: Gower, 1986), 44-45.

[76] Rom Harre, "There Is No Time Like the Present," in *Logic and Reality*, ed. B. J. Copeland (Oxford: Clarendon, 1996), 406.

interval of time may be conceptually infinitely divisible. That is to say, the dividing can go on to infinity as a limit at which one never arrives. Thus, any interval taken to be present, say, the present minute, can be subdivided into phases, which will be past, present, and future respectively. The present will be any arbitrarily chosen interval centered on a present instant. Because time is not composed of instants, temporal becoming does not proceed by instants; rather duration is conceptually prior to any sort of divisions we make in it.

Such a view is admittedly strange because it implies that there is no such thing as *the* present time. Rather what is present depends on the universe of discourse: Are we talking about seconds, or minutes, or hours, or what? And even these intervals can be analyzed into sub-intervals, not all of which are present. We instinctively feel that there must be some unique metric interval which is present absolutely, and God is sustaining it in being. But such a feeling may be the result of our mathematization of time, thinking of time on the model of a geometrical line composed of points. But it is precisely this model that such a view rejects.

None of the alternatives for understanding the extent of the present leaves one feeling entirely comfortable. But discomfort is not incoherence. It may be a reflection of how profoundly difficult time is to understand. It does not show that temporal becoming is unreal.

In fact, at this point the defender of temporal becoming may attempt to turn the tables on the static time theorist by arguing that only the dynamic view, committed as it is to the reality of temporal becoming, enables us to understand the asymmetry of time. The asymmetry of time consists in two distinct, but frequently confused, features of time: (i) the anisotropy of time and (ii) the direction of time. *Isotropy* is the property of being the same in all directions. For example, space is isotropic. There is no "up" or "down" in outer space; it is entirely arbitrary, for example, that globes and world maps of the earth always situate the North Pole on the top and the South Pole on the bottom. (A politically correct mindset might see this as indicative of the arrogance of the peoples of the northern hemisphere, always wanting to be on top and lording it over the peoples of the southern hemisphere!) But in contrast to space, time is not isotropic. It has two distinguishable directions: "earlier" and "later." There is an objective difference between being earlier than some event and being later than that event. It is virtually universally agreed that time is essentially ordered by the relations *earlier than* and *later than;* any dimension not possessing the directions "earlier" and "later" is not a temporal dimension. On the other hand, the *direction* of time has to do with time's being oriented in one direction. By contrast, the temperature scale on a thermometer is anisotropic (there is a difference between "colder" and "hotter"), but it has no inherent direction. The temperature can move in either direction. Time, on the other hand, does seem to have a direction: from past to future. If an arbitrarily chosen event occurs, and we were asked which event would occur next, we would unhesitatingly point to the event after it, not to the event before it. It is the directionality of time that gives rise to the sense of the irretrievability of the past, which comes to expression in such proverbs as, "It's no use crying over spilled milk," or, "That's water under the bridge."

Now clearly if time has a direction, then time must be anisotropic. The dynamic time theorist finds in the objective reality of temporal becoming a basis for affirming time's directionality and therefore also its anisotropy. The asymmetry of time is thus objectively grounded in temporal becoming. The static theory, on the other hand, does not seem to have any basis for affirming the asymmetry of time, so that its proponents have simply to assume time's asymmetry or to deny it.

The dynamic time theorist grounds the direction of time in the impossibility of a backward lapse of time. The absurdity of a backward lapse of time can be seen by contemplating the idea of backward continuing.[77] Temporal continuing or endurance is not simply the tenseless, temporal extension of some object. On a dynamic theory of time, the successive moments of an object's duration do not all tenselessly exist; rather they come into being and pass away. If an object exists at the present moment, then in order for it to continue to exist, another moment must come into being. But such an additional moment can only come to be *after* the present moment. It seems completely unintelligible to say that that additional moment might come to be *before* the present moment. If the moment existed before the present moment, then we would say that this was a case of the object's enduring from the past moment to the present moment. On a presentist view it simply makes no sense to say that an object continued to exist from the present moment until the past moment. Even if we imagine events occurring in reverse order, as when a film is run backwards, still the reversed-sequence events occur one after the other. What is impossible to conceive is that the moments of time themselves should occur in reverse order. Thus, on a dynamic theory time must be invested with a direction.

On a dynamic theory of time, then, the anisotropy of time and the directionality of time become perspicuous, being grounded in the nature of temporal becoming.

---

[77] See Sarah Waterlow, "Backwards Causation and Continuing," *Mind* 83 (1974): 372-387.

By contrast, the static theory of time seems to lack the resources for making time's asymmetry anything more than a groundless assumption. A great deal of ink has been spilled in the attempt to ground time's asymmetry in various physical processes such as entropy increase, the expansion of the universe, and so forth. From a theistic perspective, however, all such attempts seem misconceived. For one can easily conceive of a possible world in which God creates a universe lacking any of the typical thermodynamic, cosmological, or other arrows of time, and yet He experiences the successive states of the universe in accord with the lapse of His absolute time. There seems to be no good reason to think of the physical processes as anything more than empirical measures or indicators of the lapse of time, rather than as constitutive of the nature of time itself.

Furthermore, even considered on their own merit, such physical processes are simply *irrelevant* to a definition of temporal asymmetry. For why should we regard one direction of the physical process as the "earlier" direction rather than the "later" direction? If this decision is not to be utterly arbitrary, there must be some non-physical feature of time which serves to differentiate the "earlier" from the "later" direction. Sklar observes that we do not *mean* by "earlier than" something such as "having lower entropy than"; rather the association of lower entropy states with earlier times of a physical process is an *empirical discovery* we make about the world, a discovery which we may then use to ascertain in other cases which stages of a physical process are earlier.[78] According to Sklar, "we know independently of our knowledge of the lawlike behavior of physical processes in time, what the actual time order of events really is. Only this 'independent' knowledge of temporal order would allow us to decide which of the lawlike descriptions is, in fact, the true lawlike description of the world."[79] He makes the important observation that in the inner life of the mind I directly experience the temporal succession of experiences, and I find that the same *later than* relation also seems to characterize events in the external world. If I also discover that external events are similarly related by some physical relation, such a discovery would at best show a *correlation* but not an *identity* of the physical and temporal relations. In this sense, Sklar concludes, there really is no need for a "theory of the direction of time"—"we may suppose that at least some relations of temporal priority are also among the directly inspectable features of events."[80]

So if relations of temporal asymmetry exist and are knowable indepen-

---

[78] Sklar, *Space, Time, and Space-Time*, 403-404.
[79] Ibid., 402.
[80] Ibid., 410-411.

dently of physical processes, what is the foundation, on the static theory, of the asymmetry of time? The static time theorist seems obliged to treat time's asymmetry as just a "given." But this assumption fits ill with a static theory of time, for given the tenseless existence of all events, time seems to be isotropic and directionless. Indeed, some static time theorists such as Paul Horwich and Huw Price have therefore boldly affirmed that time is wholly symmetric.[81] Such a position is consistent with the static theory, but it seems fantastic in light of our experience of earlier and later. In order to have a credible theory of time, the static time theorist must simply assume the existence of temporal anisotropy. But that assumption seems *ad hoc* and fails to mesh naturally with his tenseless understanding of time. Thus, ironically, the so-called myth of passage, once properly analyzed, far from undermining the dynamic theory, actually redounds to its credit and serves to highlight one of its advantages over the theory of tenseless time: It furnishes a foundation for the existence of temporal asymmetry.

In conclusion of this section, we may say that neither McTaggart's Paradox nor the Myth of Passage provides good grounds for rejecting a dynamic theory of time, since these objections are in truth aimed at a hybrid, dynamic-static theory of time. So directed, they are cogent objections. But the pure dynamic time theorist, or presentist, is not at all menaced by these foes. Rather the serious difficulty he must confront is the classic problem of the extent of the present. None of the options here is without its drawbacks. The question is whether this difficulty is so great as to outbalance the arguments in favor of a dynamic theory of time.

In this chapter we have seen two powerful arguments in favor of a tensed or dynamic theory of time: the argument from the ineliminability of tense and the argument from our experience of tense. It seems to me that arguments based on the irreducibility of tensed facts and on the undeniable presentness of experience are sufficiently strong to outweigh any puzzles attending the extent of the present. The remaining question, then, is whether a consideration of arguments for and against a static theory of time will reinforce or subvert this conclusion.

Recommended Reading

The Ineliminability of Tense

Gale, Richard. *The Language of Time*. International Library of Philosophy and Scientific Method. London: Routledge & Kegan Paul, 1968.

---

[81] Paul Horwich, *Asymmetries in Time* (Cambridge, Mass.: MIT Press, 1987), 54-57; Price, *Time's Arrow and Archimedes' Point*.

Perry, John. "The Problem of the Essential Indexical." *Noûs* 13 (1979): 3-29.

Mellor, D. H. *Real Time.* Cambridge: Cambridge University Press, 1981.

Smith, Quentin. *Language and Time.* New York: Oxford University Press, 1993.

Oaklander, L. Nathan, and Quentin Smith, eds. *The New Theory of Time,* part I. New Haven, Conn.: Yale University Press, 1994.

Craig, William Lane. "Tense and the New B-Theory of Language." *Philosophy* 71 (1996): 5-26.

Recanati, François. "Direct Reference, Meaning, and Thought." *Noûs* 24 (1990): 1-18.

Craig, William Lane. "The New B-Theory's *Tu Quoque* Argument." *Synthese* 107 (1996): 249-269.

Our Experience of Tense

Prior, A. N. "Thank Goodness That's Over." *Philosophy* 34 (1959): 12-17.

Mellor, D. H. *Real Time.* Cambridge: Cambridge University Press, 1981.

———. "MacBeath's Soluble Aspirin." *Ratio* 25 (1983): 89-92.

Schlesinger, George. *Aspects of Time.* Indianapolis: Hackett, 1980.

Hestevold, H. Scott. "Passage and the Presence of Experience." *Philosophy and Phenomenological Research* 50 (1990): 537-552.

Oaklander, L. Nathan, and Quentin Smith, eds. *The New Theory of Time,* part III. New Haven, Conn.: Yale University Press, 1994.

Craig, William Lane. "The Presentness of Experience." In *Time, Creation, and World Order.* Edited by Mogens Wegener. Acta Jutlandica LXXIV: 1. Humanities Series 72. Aarhus, Denmark: Aarhus University Press, 1999.

———. "Tensed Time and Our Differential Experience of the Past and Future." *Southern Journal of Philosophy* 37 (1999): 515-537.

———. "On Wishing It Were Now Some Other Time." *Philosophy and Phenomenological Research* 62 (2001): 1-8.

McTaggart's Paradox

McTaggart, J. M. E. *The Nature of Existence,* 2 vols. Edited by C. D. Broad. Cambridge: Cambridge University Press, 1927; rep. ed.: 1968.

Wisdom, John. "Time, Fact and Substance." *Proceedings of the Aristotelian Society* 29 (1928–1929): 69-94.

Broad, C. D. *An Examination of McTaggart's Philosophy,* 2 vols. Cambridge: Cambridge University Press, 1938; rep. ed.: New York, Octagon Books, 1976.

Lowe, E. J. "The Indexical Fallacy in McTaggart's Proof of the Unreality of Time." *Mind* 96 (1987): 62-70.

Mellor, D. H. *Real Time.* Cambridge: Cambridge University Press, 1981.

Buller, David J., and Thomas R. Foster. "The New Paradox of Temporal Transience." *Philosophical Quarterly* 42 (1992): 357-366.

Le Poidevin, Robin. *Change, Cause and Contradiction: A Defense of the Tenseless Theory of Time.* London: Macmillan, 1991.

Prior, A. N. "The Notion of the Present." *Studium Generale* 23 (1970): 245-248.

Wolterstorff, Nicholas. "Can Ontology Do without Events?" *Grazer Philosophische Studien* 7/8 (1979): 177-201.

Craig, William Lane. "McTaggart's Paradox and the Problem of Temporary Intrinsics." *Analysis* 58 (1998): 122-127.

_____. "McTaggart's Paradox and Temporal Solipsism." *Australasian Journal of Philosophy* 79 (2001): 32-44.

Dummett, Michael. "A Defense of McTaggart's Proof of the Unreality of Time." *Philosophical Review* 69 (1960): 497-504.

_____. "The Reality of the Past." *Proceedings of the Aristotelian Society* 69 (1968–1969): 239-258.

Cresswell, M. J. "Modality and Mellor's McTaggart." *Studia Logica* 49 (1990): 163-170.

The Myth of Passage

Broad, C. D. *An Examination of McTaggart's Philosophy*, 2 vols. Cambridge: Cambridge University Press, 1938; rep. ed.: New York: Octagon Books, 1976.

Williams, Donald C. "The Myth of Passage." *Journal of Philosophy* 48 (1951): 457-472.

Craig, William Lane. "The Extent of the Present." *International Studies in the Philosophy of Science* 14 (2000): 165-185.

Prior, A. N. "Changes in Events and Changes in Things." In *Papers on Time and Tense*, pp. 1-14. Oxford: Clarendon, 1968.

Grünbaum, Adolf. "Relativity and the Atomicity of Becoming." *Review of Metaphysics* 4 (1950–1951): 143-186.

R. M. Blake. "The Paradox of Temporal Process." *Journal of Philosophy* 23 (1926): 645-654.

Whitrow, G. J. *The Natural Philosophy of Time*, 2d ed., chapter 4. Oxford: Oxford University Press, 1980.

Loizou, Andros. *The Reality of Time*. Brookfield, Vt.: Gower, 1986.

Waterlow, Sarah. "Backwards Causation and Continuing." *Mind* 83 (1974): 372-387.

Craig, William Lane. "Temporal Becoming and the Direction of Time." *Philosophy and Theology* 11 (1999): 349-366.

Sklar, Lawrence. *Space, Time, and Space-Time*, pp. 352-411. Berkeley: University of California Press, 1976.

Horwich, Paul. *Asymmetries in Time*, pp. 37-71. Cambridge, Mass.: MIT Press, 1987.

Price, Huw. *Time's Arrow and Archimedes' Point*. New York: Oxford University Press, 1996.

5

THE STATIC CONCEPTION OF TIME

HAVING EXAMINED THE principal arguments for and against a dynamic view of time, we now turn to a similar examination of the static conception. Although it may seem foreign to the common man, the static understanding of time is accepted almost unquestioningly by many physicists and by a good many reflective philosophers as well.

I. Arguments for a Static Conception

1. Relativity Theory

EXPOSITION

Without a doubt the paramount consideration leading people to embrace a static conception of time is Relativity Theory. It will be remembered from our earlier discussion[1] that when Albert Einstein originally formulated his Special Theory of Relativity in 1905, he presupposed a dynamic conception of time. Space and time were conceived to be separate realities—three-dimensional space enduring through the one dimension of time. But in 1908 a German mathematician by the name of Hermann Minkowski proposed that STR be understood instead in terms of a four-dimensional geometrical structure called space-time. The name "space-time" derives from the fact that three dimensions of this geometrical structure are taken to represent space and the fourth dimension represents time. The four dimensions of space-time do not differ structurally, except that the square of distances along one of the dimensions, usually taken to represent time, is negative, whereas the square of distances along the other three is positive. This is due to the fact that the four-dimensional geometry is not Euclidean. Since we are three-dimensional beings (or at least we only apprehend three dimensions), we cannot visualize what a four-dimensional object looks like. But geometers can

---
[1] See chapter 2, pages 32-66.

describe such an object mathematically, even if they cannot picture it. By treating space and time as a four-dimensional structure, mathematicians can display with great clarity the mathematical equations at the heart of STR. Such a representation reveals that while space measurements and time measurements when taken separately are relative, space-time measurements are absolute. The space-time position of events and the space-time interval between them are the same for all observers and never change.

It may seem strange to conceive of space and time as united in space-time. After all, they are so different that trying to combine them may seem like mixing oil and water. We may be inclined to think of space-time non-realistically, as a mathematical fiction which is useful in the way that diagrams and graphs are—not as realistic representations of the world, but as conceptual aids. But Minkowski was a metaphysician as well as a mathematician, and he interpreted his space-time realistically. Space-time was not merely a representation of the world of space and time; it *was* the world. Minkowski called the space-time points designated by three spatial coordinates and one temporal coordinate "world points" and the collection of all such points he christened "the world."[2] He announced a "metamorphosis of our concept of nature," and concluded with the famous words, "Henceforth, space by itself, and time by itself, are doomed to fade away into mere shadows, and only a kind of union of the two will preserve an independent reality."[3]

Minkowski's words proved to be prophetic. His space-time approach to relativity, especially after Einstein's formulation of his General Theory of Relativity (GTR), became the dominant mode of presentation of relativity. Einstein himself became an ardent space-time realist. He remarked, "Even in the relativity theory we can still use the dynamic picture if we prefer it. But we must remember that this division into time and space has no objective meaning since time is no longer 'absolute'."[4] Thus, Relativity Theory was "distinctly in favor of the static picture and found in this representation of motion as something existing in time-space a more convenient and more objective picture of reality."[5] Abandoning his original view, Einstein concluded, "It appears therefore more natural to think of physical reality as a four-dimensional existence, instead of, as hitherto, the evolution of a three-dimensional existence."[6]

---

[2] H. Minkowski, "Space and Time," in *The Principle of Relativity*, by A. Einstein, et al., trans. W. Perrett and G. B. Jeffery (New York: Dover Publications, 1952), 76.
[3] Ibid., 75, 76.
[4] Albert Einstein and Leopold Infeld, *The Evolution of Physics* (New York: Simon & Schuster, 1938), 220.
[5] Ibid., 217.
[6] Albert Einstein, *Relativity: The Special and the General Theory*, 15th ed. (New York: Crown, 1961), 150.

Construing space-time realistically implies, as Einstein's words indicate, a static conception of time. For space-time itself never changes. Change takes place *in* time, that is, along the dimension of the structure which represents time. But there is no change *of* the structure as a whole. Changes in time are like the changes in scenery from east to west. Every event in time is permanently fixed at its location. Indeed, it can be said that while space-time is *intrinsically* temporal (that is, one of its dimensions is time), it is *extrinsically* timeless (that is, it does not exist in some embedding hyper-time). Space-time neither changes nor becomes; it just *is* (tenselessly).

We thus have before us two radically different interpretations of Relativity Theory, one compatible with a dynamic view of time (the original Einstein interpretation) and one implying a static view of time (the Minkowski interpretation). Graham Nerlich, a prominent Australian philosopher of space and time, has called these "the relativity interpretation" and "the space-time interpretation" respectively.[7] These two understandings of Relativity Theory are *very* often confused, but it is crucial for our purposes that they be kept distinct. If the space-time interpretation of Minkowski is adopted, then tense and temporal becoming are squeezed out of the world as objective realities. On the other hand, while the original Einstein relativity interpretation would require us to relativize tense and temporal becoming to inertial frames (most plausibly by taking the standard simultaneity relation defined in STR[8] to pick out all events present at that time relative to that frame), such an interpretation is nonetheless compatible with a dynamic theory of time. The question is: Are there good reasons for preferring one of these interpretations over the other?

It seems that there are. The space-time interpretation is arguably superior to the relativity interpretation for three reasons. First, no plausible relation of co-existence can be defined within the context of the relativity interpretation of Einstein.[9] For any two co-existent objects A and B, A exists with B if and only if B exists with A. That is just part of the meaning of co-existence. Now given the reality of tense, two objects are co-existent if and only if they are co-present. That is because present entities are the only temporal entities which exist, given a tensed theory of time. But then how shall we understand co-existence within the context of the relativity interpretation?

Let us suppose that A and B are some distance apart and in relative

---

[7] Graham Nerlich, *What Spacetime Explains* (Cambridge: Cambridge University Press, 1994), 33.
[8] Recall our account in chapter 2, pages 39-40.
[9] See Yuri Balashov, "Enduring and Perduring Objects in Minkowski Space-Time," *Philosophical Studies* 99 (2000): 129-166.

motion, and let us imagine that an event occurs at A's location (call this event Ae) and another event occurs at B's location (call it Be). Due to the relativity of simultaneity, in A's inertial frame Be may be simultaneous with Ae and thus present for A at the time of Ae. But in B's inertial frame Ae will not be simultaneous with Be and so will be past or future for B at the time of Be. Thus, Ae and Be cannot be co-present. B is present to A, but A is not present to B. But if A and B cannot be co-present, neither can they be co-existent. Suppose we try to remedy this defect by stipulating that in order for A and B to co-exist, events in the lives of A and B must be co-present in the sense that if Be is present to Ae, then Ae must be present to Be. A and B would then be co-present and, hence, co-existent. The only problem is, on this definition no object ever co-exists with a moving object! The only objects which co-exist are objects at rest relative to each other. But this seems crazy, since virtually everything is in motion relative to myself, with the result that scarcely anything co-exists with me.

By contrast, on Minkowski's space-time interpretation, the co-existence relation can be plausibly defined because it is unconnected to the fictitious co-presence relation. Since all events in space-time are equally real, two objects A and B can be said to co-exist just in case there are events in the life of A and B which are sufficiently far apart that they cannot be connected by a light signal. If these events cannot be connected by a light signal (that is, a light signal leaving A at the time of Ae cannot get to B until after Be occurs), then that implies that in some inertial frame Ae and Be are reckoned to be simultaneous. So co-existent events will be those which can be simultaneous in some inertial frame. And A and B co-exist just in case their life-histories include such events. This account of co-existence cannot work for the relativity interpretation because only present events exist on a tensed theory of time.

Second, the relativity interpretation results in a fantastic fragmentation of reality. On the relativity interpretation, there is no unified, common world inhabited by all observers, but rather a plurality of spaces and times each associated with a different inertial frame. STR requires that even if we are merely passing each other in automobiles, our classes of simultaneous events do not coincide, and at sufficient distances various events occur and things exist in relation to me which may be future and thus literally unreal for you. But if we decelerate and come to relative rest, then we come to share the same reality; events and things which were once present and real for me are now future and unreal. Reality is relative to reference frames. One can change one's reality just by changing one's relative motion. If the relative motion between two events is great, the distance between the events need not be huge in order for the fragmentation of reality to become evident. For example, any event on the planet Neptune within a space of about eight hours can be reckoned according to STR to be occurring now by some earth observer. For one earthling Neptune could have been completely destroyed in a cosmic collision, while for another relatively moving earthling Neptune could exist perfectly well. For other observers, events on Neptune's surface literally occur in reverse order. Even at the distance of the earth's diameter, anything occurring within about one-tenth of a second could be real for us at this moment. For relatively moving observers, the Chinese president Jiang Zemin could be literally dead or alive, depending on the observers' motion. This is not a mere matter of which events are calculated to be present relative to an inertial frame. Rather reality quite literally falls apart, and there is no one way the world is.

By contrast, on the space-time interpretation, all events in space-time are equally real, and things do not pop into or out of existence as I switch reference frames. When I determine a certain class of events to be simultaneous with me-now, I am simply designating a certain slice or cross-section of space-time. A relatively moving observer using STR's method of determining simultaneous events will slice space-time at a different angle and so come up with a different class of simultaneous events than I do. There is an objective, unified world which is the same for all observers, namely, the four-dimensional space-time itself. Since reality is not tied to simultaneity, the relativity of simultaneity does not imply that reality is relative, in contrast to the relativity interpretation.

Third, the relativity interpretation is explanatorily deficient with regard to relativistic phenomena. On the relativity interpretation, physical objects are three-dimensional entities enduring through time. Yet they are said to have no intrinsic properties such as length, shape, mass, and duration. These are said to be merely relational properties—for example, having a certain length relative to a certain inertial frame. But there is no explanation or foundation for why this is so. It needs to be appreciated that, on the relativity interpretation, relativistic phenomena such as the shrinking of an object in motion or the slowing down of a moving clock are every bit as real, physical effects as they are under Lorentz's theory.[10] This seems incredible, since such effects

---

[10] See John A. Winnie, "The Twin-Rod Thought Experiment," *American Journal of Physics* 40 (1972): 1091-1094; M. F. Podlaha, "Length Contraction and Time Dilation in the Special Theory of Relativity—Real or Apparent Phenomena?" *Indian Journal of Theoretical Physics* 25 (1975): 74-75; Dieter Lorenz, "Über die Realität der FitzGerald-Lorentz Kontraction," *Zeitschrift für allgemeine Wissenschaftstheorie* 13/2 (1982): 308-312.

are reciprocal: For two relatively moving clocks A and B, A runs slow relative to B and B runs slow relative to A. But Einstein understood right from the start that such relativistic phenomena were not a matter of mere appearance, but were literal, measurable effects.[11] This is especially evident in the familiar Twin Paradox, according to which absolute effects such as differential aging ensue as a result of merely relative motion.[12]

But the relativity interpretation neither provides nor permits any causal explanation of these real, physical distortions of three-dimensional objects. These phenomena simply follow as deductions from the postulates of STR. As one commentator has remarked, "The principle of relativity of . . . Lorentz, and Poincaré resulted from careful study of a large number of experiments, and it was on the basis of a theory in which empirical data could be explained to have been *caused* by electrons interacting with an ether. Einstein's principle of relativity excluded the ether of electromagnetic theory and did not explain anything."[13] STR does not permit causal explanations of relativistic phenomena because these result from merely *relative* motion, and thus no room is left for intrinsic causal forces.

By contrast, on the space-time interpretation, three-dimensional objects do not shrink up or slow down for the simple reason that three-dimensional objects do not exist! Reality is four-dimensional, and the supposed distortion of physical objects is just a matter of looking at four-dimensional objects from different angles.[14] Just as a three-dimensional object looks foreshortened when we gaze along its length in the direction of sight, so four-dimensional objects are calculated to have different shapes depending on how they are viewed in space-time. Length contraction is just the result of applying different coordinate measures to the same, unchanging, four-dimensional object. Clocks do not literally slow down; rather the same spatio-temporal intervals are measured with different coordinate systems. Moreover, in Minkowski space-time, a curved path through space-time is actually the shortest, so that the clock of an observer following such a path will record less time than a clock following a straight path. So, for example, in the Twin Paradox, the

---

[11] A. Einstein, "Zum Ehrenfestschen Paradoxen," *Physikalische Zeitschrift* 12 (1911): 509-510.

[12] In the Twin Paradox, one twin stays at home on earth while his brother goes on a high-speed journey into outer space and back. When they meet again, relativity theory predicts that the traveling twin will have lived through a shorter time and so be younger than his stay-at-home brother. Although the story so told involves absolute motion of the traveling twin, it can be reformulated in terms of three brothers, involving only relative motion.

[13] Arthur I. Miller, "On Some Other Approaches to Electrodynamics in 1905," in *Some Strangeness in the Proportion*, ed. Harry Woolf (Reading, Mass.: Addison-Wesley, 1980), 85.

[14] See account given by Edwin F. Taylor and John Archibald Wheeler, *Spacetime Physics* (San Francisco: W. H. Freeman, 1966), 1-4.

path of the traveling twin through space-time is actually shorter than the space-time path of the twin who stays at home. Thus, it is not surprising that the traveling twin records less time and so is younger when he and his brother meet again. On the space-time interpretation, then, relativistic phenomena are not inexplicable, brute facts, but have a perspicuous foundation.

For these three reasons the space-time interpretation of STR is superior to the relativity interpretation. But if that is the case, then, as Einstein came to believe, tense and temporal becoming are illusions of human consciousness. Reality is tenseless, and the static theory of time is correct.

We can formulate this argument in favor of a static theory of time as follows:

1. Either the Einsteinian, relativity interpretation or the Minkowskian, space-time interpretation of STR is correct.

2. If the Minkowskian, space-time interpretation of STR is correct, then a static theory of time is correct.

3. The Einsteinian, relativity interpretation of STR is not correct.

4. Therefore, a static theory of time is correct.

CRITIQUE

I am persuaded that the arguments given above against the relativity interpretation are cogent and that therefore premise (3) is true. But the reader who has followed our argument thus far will realize that I consider premise (1) to be false. For that premise presents us with a false dilemma. There is a third interpretation of relativity, usually overlooked in discussions of this sort, which is empirically equivalent to the Einsteinian and Minkowskian interpretations and is fully compatible with a dynamic theory of time, namely, Lorentzian relativity.

On a Lorentzian view, there do exist absolute simultaneity and absolute length, and length contraction and clock retardation are the causal effects of absolute motion. Such an interpretation is immune to the problems afflicting the relativity interpretation. First, co-existence and co-presence are defined in terms of the absolute simultaneity of events occurring in the privileged reference frame. Everything existing at the same time in that frame is real. Second, because relations of absolute simultaneity exist, things do not pop into or out of existence as one changes inertial frames. Observers in motion who use Einstein's procedure for synchronizing clocks will calculate different distant events to be simultaneous with themselves, but the discrepancy exists only in their measurements, not in reality. Third, relativistic phenomena have real, intrinsic causes, since they result from an object's motion relative to the privileged reference frame. Clocks in motion relative to the privileged frame run slow, and measuring rods in motion shrink up.

A Lorentzian theory of relativity is wholly compatible with the reality of tense and temporal becoming, since these are characteristics of absolute time. Hypothetical observers using Einstein's conventions for synchronizing clocks may calculate that some distant event is present, past, or future depending upon their relative motion, but these judgments are not to be taken literally, since the measuring devices used by such observers are distorted in virtue of their motion relative to the privileged reference frame, and therefore their judgments are skewed. Only an observer at rest in the privileged frame can use Einstein's procedure for synchronizing clocks in order to determine what events are really present.

So long as a Lorentzian interpretation of relativity is as equally plausible as the space-time interpretation, the defender of a dynamic theory of time need not be in the least disturbed by the deficiencies in the relativity interpretation. On the contrary, he will probably see those deficiencies as quite debilitating for the relativity interpretation.

In fact, the Lorentzian may see his view as superior to the space-time interpretation precisely in view of these same considerations. First, on a Lorentzian view, absolutely simultaneous events constitute at any time a unique class of events which are co-present and co-existent. But the co-existence relation defined under the space-time interpretation is implausible. It requires us to say that two events which cannot be connected by a light signal are co-existent for some observer even if, relative to that observer, one event is in the future and the other is in the past! Remember, we are not talking about co-existence in the tenseless sense in which all events (even though connectable by a light signal) can be said to exist on the space-time interpretation. We are trying to delineate a special class of events which stand in a relation of co-existence. But it is just a misuse of words to say that, for example, I co-exist with the decay of a distant star which will not take place for another 3 billion years. Second, on a Lorentzian view things do not come to exist or cease to exist as one changes inertial frames, since things either exist or do not exist in absolute time. The space-time interpretation avoids this unwelcome consequence of the relativity interpretation only by denying that things come into being or pass away at all. This extraordinary metaphysical hypothesis not only contradicts experience but is subject to further objections which will be examined below.[15] Third, on a Lorentzian interpretation three-dimensional objects are distorted due to their absolute motion. The space-time explanation of length contraction and clock retardation requires us to hold that what seem to be three-dimensional objects are in reality but parts of four-dimensional objects, a view which is open to powerful objections.[16] Thus, the very respects in which the space-time interpretation is superior to the relativity interpretation are also those in which the Lorentzian interpretation is superior to the space-time interpretation.

If the static time theorist is to prove on the basis of STR that a tenseless theory of time is true, then he must show that the space-time interpretation of STR is superior to a Lorentzian perspective. So the question is, why should we prefer a space-time interpretation to a Lorentzian interpretation?

It is often said that Lorentzian relativity is less simple than Einsteinian or Minkowskian relativity and therefore the latter are to be preferred. But as is well-known, one cannot make a naive equation between a theory's simplicity and its truth. This is especially the case if simplicity is bought at too high a price (for example, sacrificing explanatory power or making extraordinary metaphysical commitments such as realism about space-time). In any case, it is simply false that Lorentzian theory is less simple. Although Lorentz's original theory was more complicated than Einstein's, the famous physicist H. E. Ives was able to derive the Lorentzian equations (which constitute the mathematical core of STR) from the laws of conservation of energy and momentum and from the laws of transmission of radiant energy. Ives, who was a Lorentzian, concluded, "The space and time concepts of Newton and Maxwell are retained without alteration.... It is the dimensions of the material instruments for measuring space and time that change, not space and time that are distorted."[17] On Ives's accomplishment, Martin Ruderfer observes that Ives made the same number of basic assumptions as did Einstein, so that his theory has the same "beauty," thereby elevating Lorentz's theory to the same level as Einstein's.[18] Thus, it is incorrect that simplicity favors Einsteinian-Minkowskian relativity over Lorentzian relativity.

I suspect that at the root of many physicists' aversion to Lorentzian rel-

---

[15] See pages 188-215.
[16] See pages 203-209.
[17] Herbert E. Ives, "Derivation of the Lorentz Transformations," *Philosophical Magazine* 36 (1945): 392-401; reprinted in *Speculations in Science and Technology* 2 (1979): 247, 255.
[18] Martin Ruderfer, "Introduction to Ives' 'Derivation of the Lorentz Transformations'," *Speculations in Science and Technology* 2 (1979): 243.

ativity is the conviction which comes to expression in Einstein's aphorism: "Subtle is the Lord, but malicious He is not."[19] That is to say, if there exists in nature a fundamental asymmetry, then nature will not conspire to conceal it from us. But Lorentzian relativity requires us to believe that although absolute simultaneity and length exist in the world, nature conceals these from us by slowing down our clocks and shrinking our measuring rods when we try to detect them. D'Abro voices his objection to such a conspiracy of nature:

> If Nature was blind, by what marvelous coincidence had all things been so adjusted as to conceal a velocity through the ether? And if Nature was wise, she had surely other things to attend to, more worthy of her consideration, and would scarcely be interested in hampering our feeble attempts to philosophize. In Lorentz's theory, Nature, when we read into her system all these extra-ordinary adjustments *ad hoc,* is made to appear mischievous; it was exceedingly difficult to reconcile one's self to finding such human traits in the universal plan.[20]

It must first be said that d'Abro greatly exaggerates the extent of the alleged conspiracy. After all, STR is a restricted theory: It is only uniform motion relative to the privileged reference frame which is concealed from us. But acceleration and rotation are absolute motions which nature does nothing to conceal. Moreover, as we have seen, modern equivalents of the classical aether do exist and serve to pick out a privileged reference frame; and recent experiments concerning Bell's Theorem plausibly require the existence of relations of absolute simultaneity.[21] When non-Lorentzians complain that nature is conspiring to conceal a privileged frame and absolute simultaneity from us, one wonders what evidence it would take to convince them. The harder it is for nature to provide such evidence, the less compelling the charge that she is conspiring to conceal the truth from us.

But even apart from these considerations, one must surely question the presupposition that if fundamental asymmetries exist, nature must disclose these to us. As Martin Carrier writes,

> Science would be an easy matter if the fundamental states of nature expressed themselves candidly and frankly in experience. In that case we

---

[19] A remark of Albert Einstein during a visit to Princeton, upon being informed that D. C. Miller had claimed to have detected the earth's motion through the aether (cited in Abraham Pais, *"Subtle Is the Lord . . .": The Science and Life of Albert Einstein* [Oxford: Oxford University Press, 1982], 113-114).
[20] A. d'Abro, *The Evolution of Scientific Thought,* 2d rev. ed. (1927; rep. ed.: n.p.: Dover Publications, 1950), 138.
[21] See chapter 2, pages 54-57.

could simply collect the truths lying ready before our eyes. In fact, however, nature is more reserved and shy, and its fundamental states often appear in masquerade. Put less metaphorically, there is no straightforward one-to-one correspondence between a theoretical and an empirical state. One of the reasons for the lack of such a tight connection is that distortions may enter into the relation between theory and evidence, and these distortions may alter the empirical manifestation of a theoretical state. As a result, it is in general a nontrivial task to excavate the underlying state from distorted evidence.[22]

Each sentence of Carrier's statement deserves pondering. What he says about the distortion of a theoretical state in its empirical manifestation is literally true in the case of Lorentzian relativity. If in general it is difficult to excavate the underlying state of nature from distorted evidence, if nature's fundamental states often appear in masquerade, then why is the Lorentzian account of relativistic phenomena unacceptable? Tim Maudlin, a philosopher of science who has specialized in the implications of Bell's Theorem, after surveying all the attempts to integrate the EPR results with Relativity Theory, concludes, "One way or another God has played us a nasty trick."[23] He maintains that the Lorentzian solution cannot be rejected on the grounds that it would be deceptive of nature, for the partisans of *all* the solutions say the same thing about the others. In the end, he muses, "the real challenge falls to the theologians of physics, who must justify the ways of a Deity who is, if not evil, at least extremely mischievous."[24]

As for d'Abro's complaint about finding "human traits in the universal plan," the Lorentzian might in response appeal to the so-called Anthropic Principle.[25] According to that principle, features of the universe can be seen in the correct perspective only if we keep in mind that certain features of the universe are necessary if observers like us are to exist. If the universe were not to have those features, then we would not be here to observe the ones it has. Now our very existence depends upon the maintenance of certain states of equilibrium within us. But length contraction and clock retardation are, on the Lorentzian view, the result precisely of material systems' maintaining their

---

[22] Martin Carrier, "Physical Force or Geometrical Curvature?" in *Philosophical Problems of the Internal and External Worlds,* ed. John Earman, Allen I. Janis, Gerald J. Massey, and Nicholas Rescher (Pittsburgh: University of Pittsburgh Press, 1993), 3.

[23] Tim Maudlin, *Quantum Non-Locality and Relativity,* Aristotelian Society Series 13 (Oxford: Blackwell, 1994), 241.

[24] Ibid., 242.

[25] I owe this insight to Robin Collins. For a brief explication of the Anthropic Principle, see *The History of Science and Religion in the Western Tradition: An Encyclopedia,* ed. G. B. Ferngren, E. J. Larson, and D. W. Amundsen (New York: Garland, 2000), s.v. "Anthropic Principle," by William Lane Craig.

equilibrium states while being in motion.[26] Thus, if nature lacked this compensating behavior, we would not be here to observe the fact! Given that we could not exist without it, why should we be surprised at observing nature's "conspiracy"?

But why is nature structured in such a way? Given the theistic perspective from which we approach these questions, we should hardly be surprised at discovering that the universe is designed in such a way as to support our existence. We should expect that God will have chosen laws of nature which will maintain the equilibrium states essential to our existence. Even if, as d'Abro puts it, Nature is blind, God is not; and if Nature is not wise, God is. It is not Nature, then, who is concerned with our feeble selves, who deems us worthy subjects to attend to, but the Creator and Sustainer of the universe who is mindful of man (Ps. 8:3-8). *Subtle is the Lord, merciful He is also.*

A final ostensible advantage of the space-time interpretation comes from the General Theory of Relativity. In GTR gravity is understood not as a force but in terms of the curvature of space-time. Matter is conceived to warp space-time, just as a heavy object placed on a taut rubber sheet causes a depression in the sheet. If a ball bearing is rolled across the sheet, its path will be deflected by the depression, perhaps enough that the ball bearing circles around the object and finally collides with it. In a similar way, a planet orbiting the sun is conceived to do so not because of any gravitational attraction that the sun is exerting on the planet but because the planet is, so to speak, "coasting downhill" in the curved space-time warped by the sun's mass.

Now the question raised by this geometrical approach to gravitation in GTR is whether it is to be understood realistically or merely instrumentally (that is, as a convenient tool without implications for reality). For what it is worth, most physicists are apparently content to take the theory instrumentally. Curved space-time is just a geometrical model of the force of gravity. According to the noted philosopher of science Arthur Fine, few working, knowledgeable scientists give credence to the realist construal of GTR. Rather GTR is seen as "a magnificent organizing tool" for dealing with gravitational problems: "most who actually use it think of the theory as a powerful instrument, rather than as expressing a 'big truth'."[27] It can be safely said that no scientific disadvantage arises from treating the geometrical approach to gravity as merely instrumental.

Indeed, on the contrary, it can be argued that a realist understanding of

---

[26] S. J. Prokhovnik, *Light in Einstein's Universe* (Dordrecht, Holland: D. Reidel, 1985), 84-85.

[27] Arthur Fine, *The Shaky Game: Einstein, Realism, and the Quantum Theory* (Chicago: University of Chicago Press, 1986), 123.

space-time actually obscures our understanding of nature by substituting geometry for a physical gravitational force, thus impeding progress in connecting the theory of gravity to the theory of particles. In his *Gravitation and Cosmology,* the Nobel Prize-winning physicist Steven Weinberg reflects,

> In learning general relativity, and then in teaching it to classes at Berkeley and M.I.T., I became dissatisfied with what seemed to be the usual approach to the subject. I found that in most textbooks geometric ideas were given a starring role....
>
> Of course, this *was* Einstein's point of view, and his preeminent genius necessarily shapes our understanding of the theory he created. However, I believe that the geometrical approach has driven a wedge between general relativity and the theory of elementary particles. As long as it could be hoped, as Einstein did hope, that matter would eventually be understood in geometrical terms, it made sense to give Riemannian geometry a primary role in describing the theory of gravitation. But now the passage of time has taught us not to expect that the strong, weak, and electromagnetic interactions can be understood in geometrical terms, and too great an emphasis on geometry can only obscure the deep connections between gravitation and the rest of physics.[28]

Weinberg contends that taking gravity to be a real force is "a crucial link" between GTR and particle physics, since there must then be a particle of gravitational radiation, the so-called graviton.[29] The whole search for a unified theory of the forces of nature, such as is sought in so-called super-string theory and M-theory, presupposes such a link. The geometrical approach of space-time realism is thus a positive impediment to our gaining a more integrated understanding of physics. Geometrical space-time, in Weinberg's view, should be understood "only as a mathematical tool" and "not as a fundamental basis for the theory of gravitation."[30]

In summary, while the space-time interpretation of STR is in some respects superior to the relativity interpretation, there do not seem to be comparably good reasons for preferring it to a Lorentzian approach to Relativity Theory. On the contrary, if our arguments for divine temporality are correct, then a Lorentzian theory of relativity *must* be true, since the frame coinciding with God's "now" will be privileged.

---

[28] Steven Weinberg, *Gravitation and Cosmology: Principles and Explications of the General Theory of Relativity* (New York: John Wiley & Sons, 1972), vii; cf. 147. Riemannian geometry is the geometry of a positively curved surface, such as the surface of a sphere.
[29] Ibid., 251.
[30] Ibid., viii.

Indeed, on the basis of what we have already discovered,[31] I think we have very substantive reasons to reject space-time realism. For inherent to the concept of space-time is the indissoluble unification of space and time into a four-dimensional continuum. But we have seen that time can exist independently of space. For if God, existing alone without creation, were to experience a sequence of mental events in the contents of consciousness, time would exist wholly in the absence of space. I take this simple consideration to be a knock-down argument against the view that time and space are indissolubly united in space-time. Thus, my sympathies lie with the French physicist Henri Arzeliès when he states, "The four-dimensional continuum should therefore be regarded as a useful tool, and not as a physical 'reality'."[32]

In conclusion, the superiority of the Minkowskian space-time interpretation to the original Einsteinian relativity interpretation of STR does not serve to justify the static theory of time, for this overlooks a Lorentzian approach to Relativity Theory, an approach which is at least empirically equivalent to the rival views, is no less plausible than the space-time interpretation, and yet is compatible with a dynamic theory of time. Reacting to the claim that a space-time approach to Relativity Theory shows tense and temporal becoming to be unreal, the philosopher of science Max Black is forthright:

> This picture of a "block universe," composed of a timeless web of "world-lines" in a four-dimensional space, however strongly suggested by the theory of relativity, is a piece of gratuitous metaphysics. . . . Here, as so often in the philosophy of science, a useful limitation in the form of representation is mistaken for a deficiency in the universe.[33]

So long as a Lorentzian approach to Relativity Theory is no less plausible than its competitors, the present argument for the static theory of time is unsuccessful.

2. The Mind-Dependence of Becoming

EXPOSITION

Apart from the support allegedly lent to the static conception of time by Relativity Theory, there are precious few arguments of a positive nature for a static theory of time. But in his oft-reprinted case for the mind-dependence

---

[31] See chapter 2, page 66.
[32] Henri Arzeliès, *Relativistic Kinematics,* rev. ed. (Oxford: Pergamon Press, 1966), 258.
[33] Max Black, review of *The Natural Philosophy of Time* by G. J. Whitrow, *Scientific American* 206 (April 1962), 181-182.

of temporal becoming, philosopher of science Adolf Grünbaum does present briefly three such arguments. They may serve us as the focal point for this section.

On Grünbaum's view, *being experienced* is essential to any event's occurring now and, hence, to temporal becoming. He asserts, "independently of being perceived, physical events themselves qualify at no time as occurring now and hence as such do not become."[34] "Becoming," he says, "is mind-dependent because it is not an attribute of physical events per se, but requires the occurrence of states of *conceptualized awareness*."[35]

What reasons are there to think that temporal becoming does not characterize events themselves but is a subjective phenomenon? It is at this point that Grünbaum presents his three arguments.

1. *The triviality of objective now-ness.* Grünbaum invites us to consider a statement such as "It is now 3 P.M." Such a statement is clearly informative. But if the word "now" does not refer to the content of some subjective awareness, then there seems to be nothing for it to refer to other than 3 P.M. itself. Thus the informative statement "It is now 3 P.M." becomes the trivial statement "It is 3 P.M. at 3 P.M.," which is evidently wrong-headed.

If the defender of tensed time says that "now" refers to a primitive property of now-ness or presentness, Grünbaum remains unconvinced: "I am totally at a loss to see that anything non-trivial can possibly be asserted by the claim that at 3 P.M. nowness (presentness) inheres in the events of 3 P.M. For all I am able to discern here is that the events of 3 P.M. are indeed those of 3 P.M. on the day in question!"[36]

Grünbaum's argument can be formulated in the following way:

1. "It is now 3 P.M." is an informative statement.

2. If presentness is not mind-dependent, then "It is now 3 P.M." is not an informative statement.

3. Therefore, presentness is mind-dependent.

And, of course, if presentness is mind-dependent, then it is not an objective feature of reality, as the partisans of tensed time affirm.

---
[34] Adolf Grünbaum, "The Status of Temporal Becoming," in *Modern Science and Zeno's Paradoxes* (Middletown, Conn.: Wesleyan University Press, 1967), 19.
[35] Ibid., 8.
[36] Ibid., 20.

2. *The absence of becoming from physical time.* Grünbaum considers this his most important argument. Physics knows nothing of temporal becoming. But if becoming were an objective feature of the world, then physical theories could not afford to ignore it without doing detriment to their explanatory success. Since such theories are quite successful, temporal becoming must be purely subjective.

We can formulate this argument as follows:

1. Current theories of physics take no cognizance of temporal becoming.

2. If temporal becoming is an objective feature of the world, then, if current theories of physics are explanatorily successful, they must take cognizance of temporal becoming.

3. Current theories of physics are explanatorily successful.

4. Therefore, temporal becoming is not an objective feature of the world.

But if there is no objective temporal becoming, then a static theory of time is correct.

3. *Why is it now?* Grünbaum's third argument is that a tenseless theory of time does not involve an important perplexity which besets the tensed theory of time, namely, why do the events which are now happening in 2001 become present in 2001 rather than at some other time? This is not, Grünbaum emphasizes, the same question as why the events happen in 2001. One could give a causal history leading up to the events in order to explain why the events occur in 2001. But what Grünbaum wants to know is why the events of the year 2001 become present in the year 2001 rather than sooner or later. On his view they are now in 2001 because there is some subjective awareness in the year 2001 which is apprehending the occurrence of these events at the same time as the awareness itself. But the defender of tensed time has no non-trivial answer to the question.

The argument seems to go as follows:

1. If now-ness is an objective feature of events, then there must be distinct explanations for why an event occurs in the year 2001 and why events of the year 2001 have now-ness in 2001.

2. There cannot be distinct explanations for these facts.

3. Therefore, now-ness is not an objective feature of events.

On the basis of these three arguments Grünbaum believes himself to have proved that temporal becoming is mind-dependent and, therefore, that the static theory of time is correct.

CRITIQUE

Let us consider each of Grünbaum's arguments in turn.

1. *The triviality of objective now-ness.* The defender of tensed time will want to maintain that presentness is not mind-dependent and that therefore the second premise of Grünbaum's first argument is false. "It is now 3 P.M." is an informative statement on a tensed theory of time. Indeed, it will be recalled that it was John Perry's work on "the essential indexical" which convinced philosophers that such a statement does not have the same meaning as a tenseless statement.[37] On Grünbaum's view this sentence means something like, "It is 3 P.M. simultaneous with a certain conceptualized awareness." But, as Perry showed, this is a tenseless truth which will not inform me as to whether I should leave for the meeting starting at 3 P.M. Thus, it is Grünbaum's construal of "now" which is crucially uninformative.

By contrast, in telling us that 3 P.M. is now or has presentness, as the tensed theory affirms, the statement is vitally informative. Grünbaum's mistake was that he confused *being present* with *being present at 3 P.M.* To say that 3 P.M. is present at 3 P.M. is trivial, but to say that 3 P.M. is present is informative. To say that it is now 3 P.M. is to say that of all possible times only 3 P.M. has actuality.

Thus, Grünbaum's first argument is based on a confusion and has been overtaken by subsequent developments in the philosophy of language.

2. *The absence of becoming from physical time.* A number of thinkers have challenged the first premise of Grünbaum's argument, that *current theories of physics take no cognizance of temporal becoming.* It is certainly true that now-ness plays a vital role in certain sciences such as meteorology or geology. For example, in forecasting weather or volcanic eruptions, scientists do not want to know simply the probability of a hurricane's striking Galveston or Montserrat's exploding at a certain time and date. They want to know if such events will take place next week. In other words, they want

---
[37] See chapter 4, page 118.

to know tensed facts about these events, which seems to contradict Grünbaum's first premise.

But perhaps Grünbaum would say that such concerns belong to applied science, not to the theories of physics themselves. If temporal becoming is real, it must appear in physical theory, which it does not. If this is his claim, then it is ironic that more recent advocates of tenseless time have blasted contemporary physics precisely because it is so thoroughly infected with the presumption of temporal becoming. In his book *Time's Arrow and Archimedes' Point*, Huw Price issues a ringing call for a wholesale reform of physical theory to make it truly tenseless by achieving what he calls an Archimedean perspective. According to Price, "the ordinary temporal perspective is so familiar, and so deeply imbedded, that we need to be suspicious of many of the concepts used in contemporary physics."[38] Even our regarding the Big Bang as the beginning, rather than the end, of the universe is to betray the assumption of a tensed perspective. Still more fundamentally, Price complains, "The conceptual apparatus of physics seems to be loaded with the asymmetric temporality of the ordinary world view. Notions such as *degree of freedom, potential,* and even *disposition* itself, for example, seem to embody the conception of an open future, for which present systems are variously prepared."[39] In Price's view we have only begun to imagine what physics would look like if it were thoroughly de-tensed. Grünbaum might protest that Price's concern is not with temporal becoming (which Grünbaum denies), but with temporal anisotropy (which Grünbaum affirms). But Price's point is that apart from the reality of temporal becoming, it simply becomes gratuitous to affirm the anisotropy of time, as contemporary physics does. Insofar as physical theory presupposes temporal anisotropy—which according to Price is "so very, very far"[40]—it also presupposes the objectivity of temporal becoming.

In any case we should surely call into question premise (2). Unapplied theories need take no cognizance of tense in order to be explanatorily successful. In fact, as Max Black explains, it is precisely the universal character of scientific statements that should lead us to expect that they would be stated in non-indexical, tenseless terms:

> It is easy to understand why theoretical physics should express its formal results in a language that is independent of context, using formulas or sen-

---
[38] Huw Price, *Time's Arrow and Archimedes' Point* (New York: Oxford University Press, 1996), 234.
[39] Ibid., 260.
[40] Ibid., 259.

tences, from which the occasion words are absent. This procedure has the great advantage of no reconstruction of the original context being required on the part of any reader. . . . If a scientist were to say, "I *then* saw a green flash at the edge of the sun's disk," anyone who was absent at the time of the original observation would need to know *who* spoke, and where and when, in order to obtain the intended information. No such supplementary information is needed in order to understand Boyle's law or any other freely repeatable scientific statement.[41]

This universalizing feature of scientific theories, their abstraction from the here and now, militates against capturing presentness in a scientific theory. But by making scientific theories tenseless, one need not thereby undermine their explanatory adequacy; on the contrary, one makes them applicable to all times.

The de-tensing of scientific theories has an important implication, however. It serves to underline Newton's distinction between time itself and our empirical measures thereof. Time in physics is an abstraction from what is arguably a richer metaphysical reality. For that reason all reductionistic views of time, which equate time with physical time, are bound to be inadequate. For the same reason philosophers and especially theologians cannot look to scientists to tell them about the nature of time, much less divine eternity, since physics does not even work with a full-orbed conception of time. As the philosopher Mary Cleugh has warned,

> The "t" of physics is improperly called *time*. . . . It is an abstraction from lived time, and in the process all that is distinctively temporal has been eliminated. Past, present, and future have gone: in their stead remains only the logical relation of before and after, expressed in terms of numbers. . . .[42]

Moreover, since t represents a number, mathematical operations can be carried out on it which make no sense with respect to time itself. For example, t can be assigned negative or even imaginary values! As Cleugh says, "What is the wildest absurdity of dreams is merely altering the sign for the physicist."[43] If the metaphysician can find no intelligible interpretation of such operations, he will justifiably regard them as mere mathematical tricks with no implication for reality. In a fascinating review of the time concept in var-

---

[41] Black, review of *Natural Philosophy of Time*, 181. By "occasion words" Black means indexical terms.
[42] Mary F. Cleugh, *Time and Its Importance in Modern Thought*, with a foreword by L. Susan Stebbing (London: Methuen, 1937), 46-47.
[43] Ibid., 46.

ious fields of physics alone, Carlo Rovelli has emphasized how unlike the intuitive notion of time physical time concepts are, and how diverse they are when compared among themselves.[44] He lists eight characteristics commonly associated with time:

1. One-dimensional: Time can be thought of as a collection of instants which can be arranged in a one-dimensional line.

2. Metric: Time intervals can be measured such that two intervals can be said to have equal duration.

3. Temporally global: The real variable t which we use to denote the measure of time goes through every real value from $-$infinity to $+$infinity.

4. Spatially global: The time variable t can be uniquely defined at all space points.

5. External: The flow of time is independent of the specific dynamics of the objects moving in time.

6. Unique: There are not many times, but just the time.

7. Directional: It is possible to distinguish the past from the future direction of the time-line.

8. Present: There always exists a preferred instant of time, the Now.

Rovelli then provides the following chart to illustrate the diversity of the physical time concept (Fig. 5.1). Even if one is disposed to dispute some of the specifics, there is, I think, no gainsaying Rovelli's point that physical time is very different from our ordinary notion of time and, moreover, that because time is differently defined in different fields of physics, there is no unitary notion of physical time. It is difficult to resist the conclusion that all of these operationally defined "times" are not really time at all but just var-

---

[44] Carlo Rovelli, "What Does Present Days [sic] Physics Tell Us about Time and Space?" Lecture presented at the Annual Series Lectures of the Center for Philosophy of Science of the University of Pittsburgh, 17 September 1993.

ious measures of time suitable for their respective fields of inquiry. Sklar has protested that

> If what we mean by "time" when we talk of the time order of events of the physical world has nothing to do with the meaning of "time" as meant when we talk about order in time of our experiences, why call it time at all? Why not give it an absurd name, deliberately chosen to be meaningless (like "strangeness") and so avoid the mistake of thinking that we know what we are talking about when we talk about the time order of events in the physical world?[45]

The gap between the ordinary conception of time and the physicist's "t" is so great that Black in fact did advise scientists to stop talking about "time" and to refer to their own concept simply as "t"![46] This is no doubt asking too much; but it is surely not too much to ask scientists—and philosophers as well—to quit drawing metaphysical conclusions based on the pale abstraction of time that plays a role in physics.

The notion of time used in:	has properties:
Ordinary language	1, 2, 3, 4, 5, 6, 7, 8
Thermodynamics	1, 2, 3, 4, 5, 6, 7
Newtonian mechanics	1, 2, 3, 4, 5, 6
STR	1, 2, 3, 4, 5
Cosmology	1, 2, 3, 4
GTR—proper time	1, 2, 3, 5
GTR—coordinate time	1, 3, 4
GTR—clock times	1, 2
Quantum gravity	none

Fig. 5.1: The concept of time in various fields of physics in comparison with the customary concept.

3. *Why is it "now"?* Grünbaum wants to know why an event which becomes present in 2001 becomes present in 2001 rather than at some other date. But this appears to be asking why a tautology is true. Perhaps we can interpret Grünbaum to be asking of some event why it becomes present in 2001, period—which is not a trivial question. But then why think that the

---

[45] Lawrence Sklar, "Time in Experience and in Theoretical Description of the World," in *Time's Arrow Today*, ed. Steven S. Savitt (Cambridge: Cambridge University Press, 1995), 226.
[46] Black, review of *Natural Philosophy of Time*, 182.

explanation for why the event becomes present in 2001 must be any different from the explanation for why the event tenselessly *occurs* in 2001? If the causal history leading up to an event suffices to explain why the event *occurs* in 2001, then the same causal history seems to explain why it becomes present in 2001. After all, on a tensed view of time, if an event *occurs* at *t*, then it must have presentness at *t*. When else could it possibly have presentness?

Grünbaum thinks that it trivializes the mind-independence of becoming to say that *"by definition* an event occurring at a certain clock time *t* has the unanalyzable attribute of nowness at time *t*."[47] But the defender of tensed time is not saying that this is a matter of definition. He is claiming that, necessarily, an event has presentness only when that event occurs. But if this is true—and how could it not be true?—then any explanation of why an event occurs at a certain time will explain as well why it becomes present at that time. Thus, Grünbaum's query is itself a trivial question. Any explanation of why an event occurs at *t* will also suffice to explain why that event has presentness at *t*.

In short, Grünbaum's arguments for the mind-dependence of becoming are not convincing. Given as well the failure of the appeal to STR, we have seen no good reason to think that a static conception of time is correct.

II. Arguments against a Static Conception

We now turn our attention to arguments against a static conception of time. Four objections stand out as particularly significant.

1. *"Spatializing" Time*

EXPOSITION

Proponents of a dynamic theory of time have long accused static time theorists of "spatializing" time. Milic Capek, for example, complains that "From Zeno to Russell and some contemporary misinterpretations of relativity, the fallacy of 'spatialization of time' is one of the most persistent features of our intellectual tradition."[48] Now this allegation should not be understood as the charge that the static theory of time *literally* turns time into a fourth dimension of space. After all, static time theorists affirm that time is ordered by the relations of *earlier than* and *later than,* which are uniquely temporal relations. Rather the charge of "spatializing" time is a metaphorical way of alleging that

---

[47] Grünbaum, "Status of Temporal Becoming," 27.
[48] Milic Capek, *The Concepts of Space and Time,* Boston Studies in the Philosophy of Science 22 (Dordrecht: D. Reidel, 1976), XXVI.

the static conception of time, by de-tensing time, has essentially robbed the temporal dimension of what makes it time, so that there is no justification for calling its ordering relations "earlier than" and "later than," with the result that this tenseless dimension can no longer justifiably be called "time." The Dutch philosopher of science Peter Kroes states the objection clearly: "it is not clear at all that the occurrence of events in the tenseless sense can generate a real temporal ordering. This tenseless occurrence of events only leads to a formal ordering relation between the physical events, not to a temporal ordering."[49]

In effect, then, the advocate of dynamic time is charging that the static theory of time is incoherent. For on the one hand it affirms the reality of time and temporal relations, but on the other hand it denies the reality of tense, which is foundational to time and temporal relations. We can formulate the objection:

1. If tense is not objectively real, temporal relations are not objectively real.

2. If temporal relations are not objectively real, time is not objectively real.

3. Time is objectively real.

4. Therefore, tense is objectively real.

All parties agree on premise (3). The issue depends upon what justification the dynamic time theorist can give for premises (1) and (2).

CRITIQUE

Some tenseless time theorists, such as Horwich and Price, would dispute the truth of premise (2). On their view, events in space-time are not ordered by relations of *earlier than* and *later than*. But as we have seen, that dimension called "time" in their theories is so unlike the ordinary conception of time that we are surely justified in doubting that time actually exists in their theories. Most philosophers, whether in the tenseless camp or in the tensed camp, agree that the relations *earlier than/later than* are essential to the nature of time.

So it comes down to premise (1). The challenge in justifying premise (1) is

---

[49] Peter Kroes, *Time: Its Structure and Role in Physical Theories*, Synthese Library 179 (Dordrecht: D. Reidel, 1985), 210.

overcoming the static time theorist's view that temporal relations are just unanalyzable givens. Every theory has its given assumptions. So why is the defender of static time not entitled simply to assume that temporal relations are real?

To meet this challenge, the dynamic time theorist must show how a reductive analysis of temporal relations can be given in terms of a tensed theory of time. The point goes all the way back to McTaggart. He maintained that the temporal relation *earlier than* is analyzable in terms of tenses: "The term *P* is earlier than the term *Q*, if it is ever past while *Q* is present or present while *Q* is future."[50] Remarkably, the static time theorist D. H. Mellor seems to agree.[51] Mellor actually offers three different ways of defining "earlier than" and "later than" in terms of tensed time:

Definition 1: to be earlier than = _{def.} to be more past or less future than

to be later than = _{def.} to be more future or less past than

Definition 2: *e* is earlier than *e** = _{def.} when *e* is present *e** is future, and when *e** is present *e* is past; and when *e* is present *e** is not past, and when *e* is future *e** is not present

e is later than *e** = _{def.} when *e** is present *e* is future, and when *e* is present *e** is past; and when *e** is present *e* is not past, and when *e** is future *e* is not present.

Definition 3: *e* is earlier than *e** = _{def.} *e* ceases to be future and becomes present first, and *e** ceases to be future and becomes present second; and *e* ceases to be present and becomes past first, and *e** ceases to be present and becomes past second

e is later than *e** = _{def.} *e** ceases to be future and becomes present first, and *e* ceases to be future and becomes present second; and *e** ceases to be present and becomes past first, and *e* ceases to be present and becomes past second

Michael Tooley, though himself a defender of dynamic time, contends that all such attempts to analyze temporal relations in terms of tensed concepts are viciously circular and therefore fail.[52] We therefore need to examine Mellor's three definitions more closely.

---

[50] J. M. E. McTaggart, *The Nature of Existence*, 2 vols., ed. C. D. Broad (Cambridge: Cambridge University Press, 1927; rep. ed.: 1968), 2:271.
[51] D. H. Mellor, *Real Time* (Cambridge: Cambridge University Press, 1981), 140.
[52] Michael Tooley, *Time, Tense, and Causation* (Oxford: Clarendon, 1997), chapter 6.

Definition 1 appears to give a wonderfully simple and straightforward analysis of the *earlier than* and *later than* relations in terms of tensed concepts. Nevertheless, Richard Gale has objected to a reductive analysis in terms of *more/less past* and *more/less future* because these are not "pure" tenses.[53] Gale's misgivings seem quite groundless, however. For tensed time does not consist simply of past, present, and future. There are all sorts of other tenses as well, such as the pluperfect and future perfect. And any temporal location relative to the present is a tensed time, such as "two years ago," "next Saturday," "for the last forty years," and so forth. So long as predicates such as "more past" are ascribed to an event absolutely, rather than relative to a tenseless date, they are pure tenses.

Tooley also objects to any analysis in terms of *more past* or *more future*.[54] He claims that these cannot be taken as primitive (or undefined) concepts of a tensed theory of time. But any attempt to analyze them, he continues, will involve the relations *earlier than/later than,* so that the analysis ultimately becomes circular.

But why, we may ask, can the defender of tensed time not regard tenses such as *more past* as being unanalyzable terms in his theory? The only reason Tooley gives is that *more past* is analogous to *more future,* and the concept *more future* cannot be a primitive concept because the concept *future* is not primitive. But why think that the simple tensed concept *future* cannot be taken as a primitive concept? Tooley's reasons for denying that *future* can be a primitive concept are based upon empiricist assumptions which he does not even attempt to justify. Deny those assumptions, as I think we should, and there is no reason why the theorist of tensed time should not take *more future* and *more past* as primitive concepts of his theory.

But suppose Tooley were right, that *more past* and *more future* cannot be primitive concepts. Why can we not analyze them in terms of other tensed concepts? For example, *more past* could be analyzed as *longer ago,* and *more future* could be analyzed as *further hence.* Or we could analyze them in terms of the present, a distance scale, and the orientation of time. Something is *more future,* for example, if it is at a greater distance from the present in the forward direction of time. Thus, our analysis of *more past* and *more future* need not appeal to relations of *earlier than/later than* and is, therefore, not circular.

---

[53] Richard M. Gale, *The Language of Time,* International Library of Philosophy and Scientific Method (London: Routledge & Kegan Paul, 1968), 93.
[54] Tooley, *Time, Tense, and Causation,* 163, 179-180; cf. 98-99.

In short, Mellor's Definition 1 provides an admirable reductive analysis of the temporal relations *earlier than* and *later than* in tensed terms.

What about Definition 2? This definition does not ascribe tenses to events absolutely, but relative to a time. Event e^* is future when e is present, and so forth. Both Gale and Tooley have objected that this analysis is viciously circular.[55] For to say that e^* is future at t or that e is past at t just means e^* is later than t or e is earlier than t. Thus, one is defining *earlier than/later than* in terms of *earlier than/later than*.

Now Gale and Tooley are correct that all the statements in Definition 2 are tenseless statements. But that does not automatically imply that they do not ascribe real tenses to events. In order to make such ascriptions clear, we may simply substitute different expressions which clearly do affirm the reality of tense. For example:

Definition 2': e is earlier than e^* =_{def.} there is some time t such that at t it is an objective fact that e has presentness and that e^* is future

e is later than e^* =_{def.} there is some time t such that at t it is an objective fact that e has presentness and that e^* is past

Definition 2' still consists of tenseless statements and provides an analysis relative to a time, but it clearly ascribes tenses to events. Gale and Tooley's mistake may have been the erroneous assumption that tenselessly true statements cannot be used to ascribe tenses. So far as I can see, then, Mellor's second definition also succeeds in showing how a reductive analysis of temporal relations can be provided by the advocate of tensed time.

Definition 3 differs from the other two in analyzing temporal relations in terms of temporal becoming. It seems to me that it can be considerably simplified:

Definition 3': e is earlier than e^* =_{def.} e becomes present first and e^* becomes present second

e is later than e^* =_{def.} e^* becomes present first and e becomes present second

Someone might suspect that the terms "first" and "second" are synonyms for "earlier" and "later," so that the analysis is viciously circular. But a moment's

---

[55] Gale, *Language of Time*, 90-91; Tooley, *Time, Tense, and Causation*, 161.

reflection shows that this is not the case. "First" and "second" are ordinal numbers which can be ascribed to spatial points or even abstract objects such as numbers and so are not inherently temporal. Given the order in which events become present, the temporal ordering of the events as earlier and later necessarily follows.

Thus, it seems that Mellor is quite right that if the tensed theory of time is true, then one can found temporal relations on the reality of tense or temporal becoming.

So far so good! The question that now arises is: Why think that temporal relations will still exist between events once the temporal dimension has been stripped of all tenses? It is universally admitted that no reverse reductive analysis of tenses in terms of tenseless temporal relations can be given. So why think that such relations would exist independently of tense? Again the point is McTaggart's. He held that once the temporal series of events is robbed of all tenses, it would still exist as a series, but not as a *temporal* series. It would be an atemporal series, like the natural numbers series or the letters of the alphabet. He wrote,

> it does not follow that, if we subtract the [tense] determinations from time, we shall have no series left at all. There is a series—a series of permanent relations to one another of those realities which in time are events—and it is the combination of this series with the [tense] determinations which gives time. But this other series . . . is not temporal, for it involves no change, but only an order. Events have an order. They are, let us say, in the order M, N, O, P. And they are therefore *not* in the order M, O, N, P, or O, N, M, P, or in any other possible order. But that they have this order no more implies that there is any change than the order of the letters of the alphabet, or of the Peers on the Parliament roll, implies any change. . . . It is only when change and time come in that the relations of this . . . series become relations of earlier and later, and so it becomes a [temporal] series.[56]

This tenseless series will include every entity which is a member of the temporal series, and all the members will be in the same order as they are in the temporal series. What, then, is the difference between them? Just this: The ordering relations of the temporal series are *earlier than* and *later than,* whereas the ordering relations of the atemporal series are not. What relations

---

[56] J. Ellis McTaggart, "The Unreality of Time," *Mind* 17 (1908): 461-462. McTaggart called the series of tensed times the *A*-series, the series of tenseless dates the *B*-series, and the de-tensed, atemporal series the *C*-series. I have substituted the bracketed words for clarity's sake.

do order the atemporal series? McTaggart did not think of them as spatial relations. Rather he made the ingenious suggestion:

> They are the relations "included in" and "inclusive of." Of any two terms in the [temporal] series, one is earlier than the other, which is later than the first, and by means of these relations all the terms can be arranged in one definite order. And of any two terms in the [atemporal] series, one is included in the other, which includes the first, and by means of these relations all the terms can be arranged in one definite order. And it seems to me possible . . . that it is the relations of "included in" and "inclusive of" which appear as the relations of "earlier than" and "later than". . . .[57]

On McTaggart's view, such an atemporal series when infused with tense yields a temporal series; but remove tense from time and what is left over is not a temporal series at all.

Now at this point we might imagine the protagonist of a tensed theory of time turning to the defender of static time and saying, "I have a foundation in my view of time for affirming the existence of the temporal relations *earlier than* and *later than*. But what entitles you, having stripped time of all tense, to assume that what remains is really time? Why should we regard those relations existing among members of your tenseless series as *earlier than* and *later than* rather than as some atemporal relations akin to the ordering relations *less than/greater than* which exist among members of the natural number series? Indeed, why think that any such relations exist at all? On my theory, tense entails the existence of temporal relations, and temporal relations entail the existence of tense. So why, if there really is no past, present, and future, as you claim, should we think that *earlier* and *later* still exist?"

Now what is the de-tenser to say at this point? One typical response is to deny steadfastly that there is any problem here at all. Thus Oaklander asks,

> What distinguishes *greater than* among numbers from *later than* among events? . . . the answer is not to be found in anything other than the relation itself. The temporal relation of succession is a simple and unanalyzable relation. . . . We can understand the difference between *later than* in time and "later than" (or *greater than*) in a number series because we can perceive the difference between the two relations. There is no further basis for the difference. . . .[58]

---
[57] McTaggart, *Nature of Existence*, 2:240.
[58] Nathan Oaklander, *Temporal Relations and Temporal Becoming* (Lanham, Md.: University Press of America, 1984), 17.

Now Oaklander is certainly right that we understand the difference between these two relations. But that does nothing to answer the question why we should think that in the absence of tense there would be any *earlier than* relation among events. Oaklander has little more to say than, "That's just the way it is!" But such a response would be acceptable only if the tensed time theorist could not provide a reductive analysis of temporal relations in terms of tensed concepts, or else the static time theorist could provide a reductive analysis of tensed concepts in terms of tenseless temporal relations. But the situation is not symmetric. The tensed time theorist can analyze temporal relations in terms of tense, and it is universally recognized that the reverse is not possible for the tenseless time theorist. Thus, the assumption by defenders of tenseless time that *earlier* and *later* can exist without tense appears to be gratuitous. As Mellor says, "Their 'block' universes have no more real time in them than McTaggart's does—the difference being that McTaggart sees this and they, by and large, do not."[59]

Mellor himself attempts to differentiate between *earlier* and *later* on the basis of perception and causation. He thinks that we perceive *before* and *after* among events just because our perceptions themselves are ordered as *before* and *after*. I have one perception after another; therefore, I perceive one event to follow another. Mellor believes that this rule holds for "not human senses only, but any sense able to perceive [temporal] precedence."[60] What determines the order in which we perceive events? Mellor answers that it is the causal order between my perceptions. He then claims that any causally connected pair of events can be perceived as standing in a relation of *before* and *after*. Thus, the direction of time is the direction of causation.

Mellor's account of temporal relations, however, is inadequate. Consider his basic claim that one cannot perceive the temporal order of events unless one's perceptions are similarly ordered. If a static theory of time is correct, as Mellor maintains, and God is timeless, then He perceives the temporal order among events without having perceptions which are also temporally ordered. Of course, God's perception of events is not based on physical signals (such as light beams), but Mellor himself says that

> *any* kind of event could be a perception. It is not being of some special kind—e.g. electrical or chemical or organic—that makes an event a perception. Perception is simply a causal process of acquiring belief, a process from

---

[59] D. H. Mellor, "McTaggart, Fixity, and Coming True," in *Reduction, Time, and Reality*, ed. Richard Healey (Cambridge: Cambridge University Press, 1981), 80.
[60] Mellor, *Real Time*, 145.

which no kind of event can be excluded *a priori*. . . . I am not interested only in human perception. My proposal is to apply to all perceptions of precedence, by all conceivable perceivers, among all sorts of events, things and dates, and it must be defensible as such.[61]

God cannot therefore be excluded as a perceiver of relations of temporal precedence. Yet obviously a timeless God would not have a temporally ordered series of perceptions. As Paul Helm puts it, God "knows (timelessly) the whole temporal series in rather the way in which for us certain things are known at a glance."[62] This counter-example undermines Mellor's whole account, for it shows that relations of *before* and *after* have no inherent connection to the temporal order of perceptions.

Moreover, it is far from obvious that even for temporal creatures the perception of temporal order has to do with the order of their perceptions. On a static theory of time, according to which all events in time are equally real, there seems to be no reason why causal influences cannot proceed backwards as well as forwards in time. God might know that our perception of event e_1 precedes our perception of event e_2. But if our perception of e_2 has a backward causal influence on our perception of e_1, then on Mellor's account our perception of e_2 is temporally prior to our perception of e_1—which, as God knows, is not the case.

Finally, even if we agree that causal influences all proceed in the same direction, Mellor's account falls short. For he still faces the same problem as those who try to base the arrow of time on physical processes: It is wholly arbitrary which direction one calls "earlier" and which "later." Who is to say that on the static theory of time, the direction of causation is not from later to earlier? Even more fundamentally, why think that on a static theory of time *earlier* and *later* even exist at all? The fact that all causes run in the same direction is no reason to think that this founds a temporal relation. It is hard to see how Mellor gives us anything more than a tenseless order of causation devoid of any real time.

In conclusion, I do not know of any successful attempt to prove that once time has been de-tensed, genuine temporal relations of *earlier than/later than* would still exist. Most tenseless time defenders seem content merely to stipulate that temporal relations exist on their theory. But such a stipulation is drawn into question by the dynamic time theorist's successful reduction of temporal relations to tensed facts. Given that there are many other atempo-

---

[61] Ibid., 153.
[62] Paul Helm, *Eternal God* (Oxford: Clarendon, 1988), 26.

ral relations that are analogous to the *earlier than/later than* relations, it seems incumbent upon the defender of static time to provide some justification for thinking that the relations he posits between events are truly temporal relations. In the absence of any such justification, the theorist of tenseless time does seem to stand convicted of "spatializing" time.

2. The Illusion of Becoming

We have already seen that the static time theorist's arguments that temporal becoming is merely subjective, or mind-dependent, are not sound.[63] Now we may push our inquiry a notch further by asking whether the claim that temporal becoming is mind-dependent is even coherent.

Exposition

When the defender of a static conception of time says that temporal becoming is "mind-dependent," it is not altogether clear precisely what he means. But one thing is clear: Such a claim implies that in the absence of conscious beings there would be no such thing as temporal becoming. If there were no minds, there would be no past, present, and future, things would not come into and go out of existence, the whole space-time continuum would just exist as a four-dimensional block, and change would be reduced to things' tenselessly possessing different properties at different space-time locations. This much is clear. What remains unclear, however, is how the presence of conscious minds serves to introduce temporal becoming into this static picture.

Are we to understand that there actually exists in the mental realm a temporal becoming which is absent from the physical realm? Is there a real temporal becoming of the contents of consciousness? Or are we rather to understand that temporal becoming is just as unreal in the mental realm as in the physical realm? Do mental events exist just as tenselessly as physical events?

Proponents of static time have not been very forthcoming in addressing such questions. But the advocate of dynamic time may argue that, whichever interpretation you choose, the doctrine of the mind-dependence of becoming turns out to be incoherent. His argument goes something like this:

1. The temporal becoming of mental events is either mind-dependent or it is not.

2. If it is not, then temporal becoming is objective.

---
[63] See above, 180-188.

3. If it is, then temporal becoming is objective.

4. Therefore, temporal becoming is objective.

Premise (1) takes for granted that we do experience the temporal becoming of the contents of consciousness. This is a datum of the phenomenology of temporal consciousness, as we have seen.[64] The question thus arises, is the temporal becoming of our experiences, such as the temporal becoming of physical events, mind-dependent or not? Now in one sense, the temporal becoming of mental events is obviously mind-dependent: Namely, without minds there would be no mental events at all! That is non-controversial. But we are asking whether the becoming of mental events is mind-dependent in the sense of "non-objective" or "illusory." The dynamic time theorist argues that however one answers, temporal becoming will turn out to be objective.

CRITIQUE

Consider first premise (2). Suppose the static time theorist says that the temporal becoming of our mental experiences is *not* mind-dependent. It immediately follows that temporal becoming is objective. For mental events, at least, come to be and pass away.

Perhaps the defender of static time will attempt to save the day by adopting a hybrid view: that there is no becoming in the physical world, but that there is real becoming in the mental realm of experience. Mental events do become, but physical events do not.

It is easy to show, however, that such a view leads to what Milic Capek calls an "absurd dualism" of "two altogether disparate realms whose correlation becomes completely unintelligible."[65] For example, why do I have the "now-awareness" of time t_1 instead of t_2? All the physical brain states at t_1 and t_2 never change, yet my now-awareness does change and is uniquely located. Why is there one privileged now-awareness? The defender of dynamic time has a ready answer: because only the physical states at t_1 actually exist. But for the static time theorist it is inexplicable why one now-awareness exists. Then there is the problem of temporal order. On a static view of time there is no reason why now-awarenesses should proceed in a particular order. There is no reason why my now-awareness should not leap

---

[64] See chapter 4, pages 129-143.
[65] Capek, *Concepts of Space and Time*, XLVII. The misdirected objections of Frederick Ferré, "Grünbaum on Temporal Becoming: A Critique," *International Philosophical Quarterly* 12 (1972): 426-445, become sound when directed against this dualistic view.

about willy-nilly among all the times at which I exist. Or consider the problem of the direction of time. Why does temporal becoming in the mental realm proceed in one direction only, rather than in the opposite direction or in both? Or the problem of inter-subjectivity: Why do we all experience the series of physical events in the same order and direction? Indeed, why do we share the same now? All of these questions become unanswerable on the view that mental events become while physical events do not. But such questions vanish (or are easily answered) if mental and physical events become together.

In any case, it will be of no use to the defender of divine timelessness to adopt such a hybrid view. For there will still be tensed facts and temporal becoming with respect to the mental realm, and thus the arguments for divine temporality based on the reality of tensed facts and temporal becoming go through successfully.[66]

So now consider premise (3). Suppose the static time theorist says that the becoming of mental events is itself mind-dependent. On this view mental events themselves are strung out in a temporal series and are all equally real. My now-awareness of yesterday and tomorrow is just as real as my now-awareness of today. The experience of the successive becoming of experiences is illusory. Experiences do not really come to be and pass away.

One problem of this view is that it flies in the face of the phenomenology of time consciousness. It denies that we experience the becoming of our experiences. For if we do have such an experience, then we must ask all over again whether that experience is mind-dependent or not, and so on. To halt a vicious infinite regress, the static time theorist must deny that we do experience the becoming of experiences. But such a phenomenology is patently inaccurate. As we have seen,[67] the so-called presentness of experience is a fundamental datum which the advocate of static time must account for.

Even more fundamentally, however, the position that mental becoming is illusory is incoherent. Bluntly put, even the illusion of becoming implies becoming. Becoming cannot be mere illusion or appearance because an illusion or appearance of becoming involves becoming. An idealist philosopher can consistently hold space to be illusory, for an illusion of space is not itself spatial. But an illusion of time is itself a temporal experience. A person who has a supposedly illusory experience of becoming is experiencing the becoming of his experiences, and this experiencing is itself a flux of experience. Change cannot be wholly illusory, for the illusion of change is a changing illu-

---

[66] See chapter 3, pages 88-109.
[67] See chapter 4, pages 133-136.

sion. Thus, the idea that temporal becoming is wholly illusory and unreal is self-refuting.

The static conception of time requires the mind-dependence of becoming in order to explain away our experience of time as past, present, and future and as continually becoming. Since physical reality itself neither becomes nor is tensed on such a view, becoming and tense must be purely subjective. They do not exist independently of minds. But the thesis of the mind-dependence of becoming is now seen to be deeply incoherent, since our mental lives involve becoming in the contents of consciousness. This becoming in the mental realm must be regarded as wholly illusory by the defender of static time, lest he admit the objective reality of tense and becoming and find himself saddled with an untenable dualism. But the position that becoming is wholly illusory is self-refuting, since such an illusion itself involves becoming. Thus, the static conception of time must be fatally flawed.

3. The Problem of Intrinsic Change

EXPOSITION

In our discussion of McTaggart's Paradox, we briefly touched on the problem of intrinsic change.[68] The problem posed by intrinsic change, it will be recalled, is how something can remain self-identical if it has different properties at different times.

The solution of the proponent of dynamic time is to take tense seriously and deny that any object has (present tense) different properties at different times. Since only the present time exists, an object only has those properties it presently has, and one can always make the time referred to by "the present time" short enough that the object will not experience any change of intrinsic properties during that time. On the presentist view, things exist wholly at a time and endure through time to later times. This solution to the problem of intrinsic change is known as *endurantism*.

The solution of the advocate of static time is typically to deny that things exist wholly at a time and to affirm instead that the three-dimensional objects that appear to us are in reality four-dimensional objects extended in time as well as space. The three-dimensional object we see is just a slice or part of a greater four-dimensional object. Things thus have three-dimensional spatio-temporal parts. For example, the Jimmy Carter we see is really just a part of the four-dimensional Jimmy Carter. Intrinsic change is understood as a four-dimensional object's having three-dimensional parts with different

---

[68] See chapter 4, pages 148-149; cf. page 87.

properties. Thus, Jimmy Carter today is not the same man who once was president. For both of them are parts of the four-dimensional Carter, and they obviously are not the same part. Thus, the spatio-temporal parts of a four-dimensional object are not identical, since they are different parts, and so they can have different properties. But the overall four-dimensional object never undergoes intrinsic change. It just has parts with different properties. Just as a three-dimensional object can be round at one end and flat at another end, so a four-dimensional object can have parts which differ in their properties. Thus, neither four-dimensional objects nor their parts endure through time, since time is one of their dimensions. In order to characterize the way in which four-dimensional objects are extended in time, philosophers have said that such objects "perdure." Accordingly, this solution to intrinsic change is called *perdurantism*.

Perdurantism obviously presupposes a static view of time. But must a defender of static time be a perdurantist? Could he embrace space-time realism and yet conceive of objects as enduring through time, having spatial, but not spatio-temporal, parts?

It seems that endurantism is not an option for the static time theorist.[69] For if he embraces endurantism, he is left without any viable solution to the problem of intrinsic change. He can no longer hold that intrinsic change is an object's different spatio-temporal parts' possessing different properties, for endurantism denies that objects have spatio-temporal parts. An object existing at t_1 and t_2 is the same object, not two parts of a greater, extended superobject. But how can it be the *identical* object at t_1 and t_2 if it has different properties at t_1 and t_2? For two objects to be identical, they must have all the same properties. The dynamic time theorist escapes the problem because he denies that t_1 and t_2 both exist. If t_1 is present, then the only object which exists is the object at t_1. But the static time theorist is stuck with the equal reality of both t_1 and t_2 and hence with the objects existing at those times. Since the objects at those respective times are not identical, it follows that things do not endure after all but only exist for an instant; change is an illusion. Thus, the static time theorist, if he is to affirm the reality of intrinsic change, must embrace perdurantism.[70]

Perdurantism is, however, a controversial doctrine beset with difficulties. Thus, the dynamic time theorist may argue:

---

[69] For a good discussion see William R. Carter and Scott Hestevold, "On Passage and Persistence," *American Philosophical Quarterly* 31 (1994): 269-283.

[70] Some static time theorists have attempted to find an alternative in what is called *adverbialism*. For an effective critique of this doctrine, see Trenton Merricks, "Endurance and Indiscernibility," *Journal of Philosophy* 91 (1994): 165-184.

1. If a static conception of time is correct, then the doctrine of perdurantism is true.

2. The doctrine of perdurantism is not true.

3. Therefore, the static conception of time is not correct.

We have already seen the justification for premise (1). It remains to be seen what objections can be raised to the doctrine of perdurantism.

CRITIQUE

Some philosophers doubt whether the doctrine of perdurantism can even be coherently formulated. There is a real danger of circularity: A perduring object is defined as a collection of spatio-temporal parts; but a spatio-temporal part is then defined as a piece of a perduring object. Because these definitions are circular, they give us no understanding of what a perduring object or a spatio-temporal part really is. This problem has driven some perdurantists to rather desperate lengths. Mark Heller, a leading perdurantist who teaches at Southern Methodist University, for example, defines a physical object as simply the material content of a region of space-time.[71] But this is a bizarre view of physical objects, for it entails that the material content of the region of space-time occupied by some of my left arm and shirt sleeve, some of my desktop, and a quantity of the adjacent air constitutes a physical object! If we say that the boundaries of a four-dimensional object are fixed by the spatial boundaries of its three-dimensional parts, then we face the difficult question of which three-dimensional objects go to comprise the four-dimensional whole. Remember that, on perdurantism, we are not dealing with the same three-dimensional object which is enduring through time; rather we are collecting together into a four-dimensional whole entirely distinct and diverse objects—but then how do we know which ones combine to form a perduring whole? Since all the three-dimensional objects are non-identical, it seems to become arbitrary. This situation has led some perdurantists such as Michael Jubien to advocate the even more radical doctrine of *object conventionalism:* the view that no physical objects at all objectively exist. According to Jubien, there really are no things—there is just "stuff" spread around in space-time, and we carve it up according to human conventions into different objects.[72]

---

[71] Mark Heller, "Temporal Parts of Four-Dimensional Objects," *Philosophical Studies* 46 (1984): 325.
[72] Michael Jubien, *Ontology, Modality, and the Fallacy of Reference* (Cambridge: Cambridge University Press, 1993), 1.

Heller, too, came to embrace conventionalism, even affirming that the objects we call "people" are merely conventional.[73]

Conventionalism is an extraordinarily high price to pay for perdurantism. Heller himself confesses to being somewhat embarrassed by the "jokes people cannot help making when they are confronted with someone who claims that there are no people."[74] Even worse, conventionalism seems to be incoherent. For, according to Heller, conventionalism affirms that (i) conventions are made by people, (ii) people are themselves conventional, and (iii) conventional objects do not really exist![75] Heller tries to elude this incoherence by admitting that many of his own statements are, strictly speaking, false; but he does nothing, so far as I can see, to modify statements (i)-(iii) or to remove the incoherence. Thus conventionalism seems a counsel of despair.

Perhaps these problems of formulating a coherent doctrine of perdurantism can be overcome. We may leave that task to the perdurantists. For perdurantism faces formidable objections on quite other grounds.

1. *Perdurantism's account of intrinsic change is implausible.* Perdurantism is the static time theorist's solution to the problem of intrinsic change. But it is evident that according to perdurantism there just is no intrinsic change. The spatio-temporal parts of an object never change, nor does the hyper-object itself change. Change is just different parts' having different properties. Since these parts are themselves distinct objects, no thing ever changes. Thus change is really an illusion.

This is a strange doctrine, to say the least. It implies that when an object appears to change, what we are really seeing is a succession of quite distinct and different objects one after another. One critic has reacted to this by saying,

> this seems to me a crazy metaphysic. . . . if I have had exactly one bit of chalk in my hand for the last hour, then there is something in my hand . . . which was not in my hand three minutes ago, and indeed, [is] such that no part of it was in my hand three minutes ago. As I hold the bit of chalk in my hand, new stuff, new chalk keeps constantly coming into existence *ex nihilo*. That strikes me as obviously false.[76]

---

[73] Mark Heller, *The Ontology of Physical Objects: Four-Dimensional Hunks of Matter* (Cambridge: Cambridge University Press, 1990), 23.
[74] Ibid., 111.
[75] Ibid., 66.
[76] Judith Jarvis Thomsen, "Parthood and Identity over Time," *Journal of Philosophy* 80 (1983): 213.

Heller tries to mitigate the apparent craziness of perdurantism by denying that the new chalk comes into being *ex nihilo* (out of nothing).[77] On Heller's view, the whole space-time region occupied by the chalk exists, and we just experience portions of it successively. This response helps, I think, to allay the charge of *ex nihilo* creation. But it does nothing to qualify the conclusion that what I hold in my hand is a wholly different object than what I was holding a second ago. The shorter piece of chalk I now hold is not the remainder of a longer piece I held before; they are utterly discrete objects. This certainly seems implausible.

Heller also attempts to turn the tables on the endurantist. He takes as his springboard Notre Dame philosopher Peter van Inwagen's critique of the so-called Doctrine of Arbitrary Undetached Parts.[78] This doctrine states that if a region of space is occupied by a material object, then the material content of *any* sub-region of that space also constitutes a material object. This doctrine would imply that the matter comprising my big toe, my right side, and a strip in between is a material object (just what Heller thinks, as we have seen!). Van Inwagen presents an ingenious argument against the Doctrine of Arbitrary Undetached Parts. He says the doctrine must be false because it necessarily implies that it is impossible for any material object to lose one of its parts, which is obviously false. He argues as follows: Imagine some object which has a part not vital to its continued existence (for example, me and one of my kidneys). According to the Doctrine of Arbitrary Undetached Parts, there exists in this case another object in addition to myself which is comprised of me minus my kidney. Now the whole object is obviously not identical to the supposed object which is the whole object minus the part. For the whole object and this diminished object lacking the part do not have exactly the same size or constituents. Suppose, then, that the whole object actually loses the part (I donate one kidney for transplantation). Since the part is not essential to the object's continued existence, the object still exists after losing the part (I survive the operation). Now the original object *is* identical to the diminished object. But this scenario violates the Principle of the Transitivity of Identity: If $x=y=z$, then $x=z$. The best way out of this muddle, van Inwagen advises, is to deny the Doctrine of Arbitrary Undetached Parts. Before the object loses its part, there just is no such object as the diminished object. There is the whole object, and the alleged diminished object is a figment of the imagination.

---

[77] Heller, "Temporal Parts," 332.
[78] See Peter van Inwagen, "The Doctrine of Arbitrary Undetached Parts," *Pacific Philosophical Quarterly* 62 (1981): 123-126.

Heller thinks van Inwagen's solution to the problem is implausible. Instead he maintains that once an object has lost its part, it is no longer identical to the original object. There are two objects—the whole object and the diminished object—and the whole object just ceases to exist. Heller's solution just is a rejection of endurance. On his view I do not survive the kidney transplant operation. There is no intrinsic change of objects. Since he regards the endurantist denial of the Doctrine of Arbitrary Undetached Parts as implausible, he takes this as evidence for perdurantism.

But Heller's reasoning is surely perverse! It seems far more obvious that one can (and does) survive a kidney transplant operation than that arbitrary undetached parts exist. Heller's perdurantism entails the Doctrine of Arbitrary Undetached Parts. Such a view not only denies the reality of intrinsic change but also results in the existence of all sorts of pseudo-objects, as we have seen. Thus, in my mind, far from making perdurantism more plausible, the warranting of such pseudo-objects makes it all the more incredible. At the very least, the endurantist denial of the Doctrine of Arbitrary Undetached Parts hardly counts against endurantism, as Heller thinks.

Thus, it seems to me that in denying the reality of intrinsic change, perdurantism presents us with an implausible view of change.

2. *Perdurantism flies in the face of the phenomenology of personal consciousness.* One of the strangest features of perdurantism is its account of persons and personal identity. On the perdurantist view, persons are not what we normally think them to be: self-conscious individuals who act and react with other things in space and time. Such individuals are, on the perdurantist view, just spatio-temporal parts or stages of persons, which are really four-dimensional objects. As such, persons are not self-conscious and have no intelligence, no volition, no emotions, no interactions, no agency, no moral responsibility, no aesthetic appreciation, indeed, virtually none of the properties we normally associate with persons. Persons, on this view, are four-dimensional objects which are changeless, non-conscious entities.

But surely such a view is absurd. I cannot imagine any sane individual who would say, if asked, that he or she is not a person. In the phenomenon of self-consciousness we immediately know ourselves as persons. Why would anyone want to embrace so outlandish a solution to the problem of intrinsic change as one that forces me to deny that I, as a self-conscious individual, am a person?

Now perhaps the perdurantist could revise his view such that we are in fact persons, and the four-dimensional object is a hyper-person or a meta-person or what have you. But such a revision would cause intractable problems for personal identity. Since spatio-temporal parts are themselves distinct objects, it follows that a meta-person is composed of a series of distinct persons! But am I seriously to believe that I at this second am not the same person as the one who was here a second ago, that my memories are really recalling some other person's experiences, that my consciousness of personal continuity from one moment to the next is an illusion? It would seem crazy to believe such things.

Consider once more the problem of intrinsic change. Perdurantism denies that any object goes through intrinsic change. But I have every reason to believe that there is at least one thing which endures through intrinsic change, namely, I myself. I existed a second ago, and despite the myriad changes which have taken place in me I still exist now. Endurantism, by taking tense seriously, permits the preservation of personal identity over time. But perdurantism forces us to say that I did not exist one second ago, nor shall I endure for another moment.

3. *Perdurantism is incompatible with moral responsibility, praise, and blame.* Since a person is conceived to be a non-conscious, four-dimensional object, it becomes nonsensical to treat such an object as a moral agent. The perdurantist might try to avoid this unwelcome conclusion by insisting that the spatio-temporal parts or stages of persons are moral agents. But then it becomes impossible to hold one person-stage responsible for what another person-stage has done. How can one person-stage be blamed and punished for what an entirely distinct, different person-stage did? Why should I be punished for his crimes? By the same token, how can moral praise be given to a person-stage for what some other, no longer existent person-stage did? Why should I, who have done nothing, get the credit for the heroism of some other person-stage?

Since moral responsibility is incompatible with perdurantism and we have more reason for affirming the former than the latter, we ought to reject perdurantism. This argument should carry special weight for the theist, for he affirms not only that people are responsible moral agents but also that God is just in holding them responsible and is Himself a virtuous agent who is to be praised for His gracious deeds.

4. *Perdurantism implies an implausible view of essential properties.* Again the ingenious van Inwagen has developed a powerful argument against perdurantism which is analogous to his objection to the Doctrine of Arbitrary Undetached Parts.[79] He invites us to consider the French philosopher

---

[79] Ibid., 133-137; idem, "Four-Dimensional Objects," *Noûs* 24 (1990): 252-254.

Descartes as a four-dimensional object which perdured from 1596 to 1650. Descartes's temporal extent is not essential to him: He might have died younger or lived longer. In that case Descartes would have been composed of a different set of temporal parts, a set formed by subtracting or adding some parts from or to the set of parts he did have. Let us call the temporal part of Descartes which is all of him except for the last year of his life "Diminished-Descartes." On the perdurantist view Descartes and Diminished-Descartes are not identical. After all, they have different temporal extents and different spatio-temporal parts. Now Descartes could have died a year earlier than he did and thus could have lacked the temporal part which is the last year of his life. But if Descartes had died one year earlier than he did, then Descartes and Diminished-Descartes would have been identical. But if Descartes and Diminished-Descartes could have been identical, then there exist two non-identical things which could have been the same thing. But this violates both the Principle of the Transitivity of Identity, as well as the Principle of the Necessity of Identity (if $x=y$, then *necessarily* $x=y$). Hence, Descartes is not a perduring object composed of temporal parts.

A simpler way of proving the same point is to consider the largest temporal part of Descartes, which just is the whole four-dimensional object called Descartes. Temporal parts have their extents essentially (an hour, for example, could not be any shorter or longer and still be an hour). So Descartes had his temporal extent essentially. But that implies that Descartes could not have lived any longer or shorter than he did—which is obviously false.

What can the perdurantist say in response to this argument? Heller's reaction has been inconsistent. At first he bit the bullet and affirmed that because spatio-temporal objects have their boundaries essentially, Descartes could not have died one year before he did.[80] This seems wildly implausible, but Heller insists that we have no choice, since endurantism is demonstrably wrong. His argument against endurantism is that it is uniquely vulnerable to the ancient Greek paradoxes known as sorites. These were puzzles typically dealing with relations between a whole and its parts. For example, how many bits of an object may be removed before it ceases to be that object? Heller maintains that endurantism can give no good answer to such sorites paradoxes.

The problem with this response by Heller is that the same sorites paradoxes can be applied to a four-dimensional object as to a three-dimensional object. How many spatio-temporal bits could be removed from a four-dimensional object before it would cease to be that object? The way Heller avoids

---
[80] Heller, *Ontology of Physical Objects*, 28.

that problem is by affirming that four-dimensional objects have their spatio-temporal boundaries essentially. But the three-dimensionalist could with equal justification say that three-dimensional objects have their spatial boundaries essentially. Thus, it is not perdurantism which does the work for Heller in avoiding the sorites paradoxes, but a quite distinct doctrine which is independent of the number and kind of dimensions a thing has.[81] In any case, there are also available less radical endurantist solutions to the sorites paradoxes.[82]

In a later response, Heller denies that according to perdurantism a thing has its spatio-temporal boundaries essentially.[83] He now maintains that a thing's identity can be fixed, not by its boundaries but by some individuating principle of unity. Thus a thing could have different spatio-temporal boundaries and remain the same thing. The idea is that Descartes could have had a different temporal extent than he has and yet still be Descartes.

On this revised view, a temporal part is not identified by its extent but by some other principle. For example, if Philip has been drunk, then we may speak of his drunken spatio-temporal part. If he had drunk less, his drunken part would have been shorter. That does not mean that the hours during which he was drunk would have contracted down to, say, only thirty minutes each. Rather it means, on Heller's view, that Philip's drunken spatio-temporal part could have been shorter because its identity is determined by its drunkenness, not by its length.

The problem for Heller's view becomes evident when we consider a statement such as, "If Philip had drunk less, then his drunken part would not have been drunk." This statement seems to be true. The spatio-temporal part of Philip which is drunk could have not been drunk. But on Heller's view that statement must be false, for drunkenness is one of the identity conditions of Philip's drunken part and thus essential to it. So Heller has to say that if Philip had drunk less, his drunken part would not have existed. But that seems bizarre. Is it not more plausible that his drunken part would have been sober? Compare a spatial analogy. Suppose we say, "If Philip had worn a hat, his sunburned face would not have been sunburned." We naturally take this to mean that his sunburned part might lack the property of being sunburned. But it seems crazy to say that this means that if he had worn a hat, Philip would have lacked one of his spatial parts, namely, his face. That would be to affirm that

---

[81] The name of this doctrine is mereological essentialism.
[82] See David S. Oderberg, *The Metaphysics of Identity over Time* (New York: St. Martin's Press, 1993), 166-173; Peter van Inwagen, *Material Beings* (Ithaca, N.Y.: Cornell University Press, 1990), 91-105, 228-229.
[83] Mark Heller, "Varieties of Four-Dimensionalism," *Australasian Journal of Philosophy* 71 (1993): 50-51.

Philip's face cannot be either pale or sunburned. Being pale or sunburned is doubtless an accidental, not an essential, property of Philip's face. In the same way, drunkenness is not an essential property of Philip's spatio-temporal parts and so does not belong to their identity conditions but is rather an accidental property of them. In short, temporal parts are to be identified by their extents, which fix their boundaries. But then we are stuck with the implausible view that Descartes could not have lived any longer or shorter than he did.

Jubien's response to van Inwagen is even more desperate. As an object conventionalist, he would escape the argument by denying that there is any such thing referred to by the name "Descartes."[84] On Jubien's view, to say that something is Descartes is not to make an identity statement at all. Rather it is to assert that some "stuff" in a certain region of space-time has the property *being Descartes*. The fact that Descartes could have lived longer or shorter is interpreted to mean that other amounts of stuff could have had the property *being Descartes*. Thus, there is no object denoted by "Descartes" at all; there is merely some stuff with the property *being Descartes*, and some other quantity of stuff might have had that property instead.

Such a view of reference and identity will hardly commend itself to most philosophers as more plausible than the premises of van Inwagen's argument. It requires us to say that Descartes as such does not exist. Not even the four-dimensional chunk of stuff which has the property *being Descartes* is Descartes. But that seems crazy. How could something have the property of *being Descartes* without being Descartes? Moreover, this view requires us to hold that identity statements are really disguised ascriptions of properties. But identity statements are necessary, whereas Jubien's property ascriptions are not. Furthermore, it requires us to say that anything could have had the property of *being Descartes*. But is this possible? Could I have actually been Descartes? Could my cat Muff? How could anyone possibly possess the property of *being Descartes* unless he is Descartes? But then such property ascriptions presuppose identity statements.

The fact that perdurantists are driven to such desperate expedients to avoid van Inwagen's argument merely testifies to its soundness and the plausibility of its conclusion. Objects do not perdure.

For all these reasons, perdurantism is an extremely implausible doctrine—certainly, at least, less plausible than its competitor, endurantism. But since a static theory of time implies the doctrine of perdurantism, the static

---

[84] Jubien, *Ontology*, 35-36.

conception of time is therefore as equally implausible, wholly apart from its other problems.

4. Creatio ex Nihilo

EXPOSITION

Christian thinkers must assess any position not only philosophically and scientifically but also theologically. A position which is philosophically and scientifically tenable but which is theologically incompatible with Christian doctrine is thereby exposed as false. Thus theological objections to metaphysical worldviews must be taken very seriously.

The static conception of time does seem to be theologically problematic in that it significantly compromises the biblical doctrine of *creatio ex nihilo* (creation out of nothing). Accordingly, the proponent of a dynamic conception of time may argue:

1. If the static conception of time is correct, a robust doctrine of *creatio ex nihilo* is not true.

2. A robust doctrine of *creatio ex nihilo* is true.

3. Therefore, the static conception of time is not correct.

CRITIQUE

The Christian theist is committed to a robust doctrine of *creatio ex nihilo* and so accepts premise (2). The Bible begins with the words, "In the beginning God created the heavens and the earth" (Gen. 1:1). Thus with majestic simplicity the author of the opening chapter of Genesis differentiated his viewpoint not only from that of the ancient creation myths of Israel's neighbors but also effectively from pantheism, panentheism, and polytheism. For the author of Genesis 1, no pre-existent material seems to be assumed, no warring gods or primordial dragons are present—only God, who is said to "create" (*bārā'*, a word used only with God as its subject and which does not presuppose a material substratum) "the heavens and the earth" (*'ēṯ haššāmāyim we 'ēṯ hā 'āreṣ*, a Hebrew expression for the totality of the world or, more simply, the universe). Moreover, this act of creation took place "in the beginning" (*berē'šîṯ*, used here as in Isa. 46:10 to indicate an absolute beginning). The author thereby gives us to understand that the universe had a temporal origin and thus implies *creatio ex nihilo* in the temporal sense that

God brought the universe into being without a material cause at some point in the finite past.[85]

Later biblical authors so understood the Genesis account of creation.[86] The doctrine of *creatio ex nihilo* is also implied in various places in early extra-biblical Jewish literature.[87] And the Church Fathers, while heavily influenced by Greek thought, dug in their heels concerning the doctrine of creation, sturdily insisting, with few exceptions, on the temporal creation of the universe *ex nihilo* in opposition to the eternity of matter.[88] A tradition of robust argumentation against the past eternity of the world and in favor of *creatio ex nihilo,* issuing from the Alexandrian Christian theologian John Philoponus, continued for centuries in Islamic, Jewish, and Christian thought.[89] In 1215, the Catholic Church promulgated temporal *creatio ex nihilo* as official church doctrine at the Fourth Lateran Council, declaring God to be "Creator of all things, visible and invisible, . . . who, by His almighty power, from the beginning of time has created both orders in the same way out of nothing." This remarkable declaration not only affirms that God created everything *extra se* without any material cause, but even that time itself had a beginning. The doctrine of creation is thus inherently bound up with temporal considerations and entails that God brought the universe into being at some point in the past without any antecedent or contemporaneous material cause.

Unfortunately, many contemporary theologians evince an unseemly timorousness concerning the biblical affirmation of *creatio ex nihilo*. Claiming that "creation is concerned with ontological origin, not temporal beginning,"[90] John Polkinghorne states, "The doctrine of creation is not an asser-

---

[85] On Genesis 1:1 as an independent clause which is not a mere chapter title, see Claus Westermann, *Genesis 1–11,* trans. John Scullion (Minneapolis: Augsburg, 1984), 97; John Sailhamer, *Genesis,* Expositor's Bible Commentary 2, ed. Frank Gaebelein (Grand Rapids, Mich.: Zondervan, 1990), 21.

[86] See, e.g., Prov. 8:27-29; cf. Ps. 104:5-9; also Isa. 44:24; 45:18, 24; Ps. 33:9; 90:2; John 1:1-3; Rom. 4:17; 11:36; 1 Cor. 8:6; Col. 1:16, 17; Heb. 1:2-3; 11:3; Rev. 4:11.

[87] E.g., 2 Maccabees 7:28; 1QS 3:15; Joseph and Aseneth 12:1-3; 2 Enoch 25:1ff; 26:1; Odes of Solomon 16:18-19; 2 Baruch 21:4. For discussion, see Paul Copan, "Is *Creatio ex nihilo* a Post-biblical Invention?: An Examination of Gerhard May's Proposal," *Trinity Journal* 17 (1996): 77-93.

[88] *Creatio ex nihilo* is affirmed in the *Shepherd of Hermas* 1.6; 26.1 and the *Apostolic Constitutions* 8.12.6, 8; and by Tatian *Oratio ad graecos* 5.3; cf.4.1ff; 12.1; Theophilus *Ad Autolycum* 1.4; 2.4, 10, 13; and Irenaeus *Adversus haeresis* 3.10.3. For discussion, see Gerhard May, *Creatio ex nihilo*: The Doctrine of "Creation Out of Nothing" in Early Christian Thought, trans. A. S. Worrall (Edinburgh: T. & T. Clark, 1994); cf. Copan's review article in note 87, above.

[89] See Richard Sorabji, *Time, Creation and the Continuum* (Ithaca, N.Y.: Cornell University Press, 1983), 193-252; H. A. Wolfson, "Patristic Arguments against the Eternity of the World," *Harvard Theological Review* 59 (1966): 354-367; idem, *The Philosophy of the Kalam* (Cambridge, Mass.: Harvard University Press, 1976); H. A. Davidson, *Proofs for Eternity, Creation, and the Existence of God in Medieval Islamic and Jewish Philosophy* (New York: Oxford University Press, 1987); Richard C. Dales, *Medieval Discussions of the Eternity of the World,* Studies in Intellectual History 18 (Leiden: E. J. Brill, 1990).

[90] John Polkinghorne, critical notice of *Cosmos as Creation,* ed. Ted Peters, *Expository Times* 101 (1990): 317. According to Polkinghorne, "To speak of God as Creator is not to attempt an answer to the question Who lit the blue touch paper of the Big Bang? To talk in that way belongs to deism and not to Christian

tion about what God did in the past to set things going; it is an assertion of what he is doing in the present to maintain the universe in being."[91] In fact, however, nearly the opposite is the case, biblically speaking. Creation in the Bible virtually always involves the notion of a temporal beginning (as is evident simply from the ubiquitous past-tense, rather than present-tense, verbs with respect to God's creating), and one will have to search hard for passages supporting the notion of the ongoing ontological dependence of the universe upon God's sustaining will. Those passages are there to be found (Heb. 1:3); but we are everywhere confronted with the idea that at some point in the past God created the world. After surveying the data, George Hendry concludes that "Creation in the language of the Bible unquestionably connotes origination . . . , the bringing into existence of something that did not previously exist."[92] A robust doctrine of creation therefore involves both the affirmation that God brought the universe into being out of nothing at some moment in the finite past and the affirmation that He thereafter sustains it in being moment by moment.[93]

Now the static time theorist can ingenuously make only the second affirmation. For him *creatio ex nihilo* means only that the world depends immediately upon God for its existence at every moment. The static time theorist's affirmation that God brought the universe into being out of nothing at some moment in the finite past can at best mean that there is (tenselessly) a moment which is separated from any other moment by a finite interval of time and before which no moment of comparable duration exists and that whatever exists at any moment, including the moments themselves, is tenselessly sus-

---

theology" (John Polkinghorne, "Cosmology and Creation," address at Trinity Hall, Cambridge, undated photocopy). "There is general agreement that the Big Bang is nothing special from a theological point of view. . . . The idea of *creatio ex nihilo* asserts the total dependence of the universe upon the sustaining will of its Creator" (Polkinghorne, critical notice of *Cosmos as Creation*, 317). Whether the Big Bang represents the moment of creation is, however, irrelevant to the conceptual content of the doctrine of *creatio ex nihilo*. The biblical doctrine, like deism, affirms a temporal beginning of the universe; moreover, deists did not in fact deny God's conservation of the world in being but rather His supernatural action in the world.

[91] Polkinghorne, "Cosmology and Creation." So also Langdon Gilkey, *Maker of Heaven and Earth* (Garden City, N. Y.: Doubleday, 1959); Ian Barbour, *Issues in Science and Religion* (New York: Harper & Row, 1971), 384; Arthur Peacocke, *Creation and the World of Science* (Oxford: Clarendon, 1979), 78-79. This watered-down doctrine of creation is the legacy of the father of modern theology F. D. E. Schleiermacher. See his *The Christian Faith*, 2d ed., ed. H. R. Mackintosh and J. S. Stewart (Edinburgh: T. & T. Clark, 1928), sec. 36-41. While acknowledging that the biblical conception of creation involves a temporal beginning (sec. 36.2), Schleiermacher held that this component of the doctrine could be safely suppressed in favor of the absolute dependence of the creation on God (sec. 41). See remarks by Nelson Pike, *God and Timelessness*, Studies in Ethics and the Philosophy of Religion (New York: Schocken, 1970), 107-110.

[92] George S. Hendry, "Eclipse of Creation," *Theology Today* 28 (1972): 420. So also Copan, "Is *Creatio ex nihilo* a Post-biblical Invention?" 77-93.

[93] For more on this distinction, see William Lane Craig, "Creation and Conservation Once More," *Religious Studies* 34 (1998): 177-188.

tained in being immediately by God. All this adds to the doctrine of ontological dependence is that the tenselessly existing block universe has a front edge. It has a beginning only in the sense that a yardstick has a beginning. There is in the actual world no state of affairs of God existing alone without the space-time universe. God never really brings the universe into being; as a whole it co-exists timelessly with Him.

Leftow, whose theory of divine eternity entails a static theory of time, admits as much. He writes,

> So if God is timeless and a world or time exists, there is no phase of His life during which He is without a world or time or has not yet decided to create them, even if the world or time had a beginning.
>
> ... God need not *begin* to do anything, then, in order to create a world with a beginning. That action that from temporal perspectives is God's beginning time and the universe is in eternity just the timeless obtaining of a causal dependence or sustaining relation between God and a world whose time has a first moment.
>
> ... in eternity, God is changelessly the Lord: He timelessly coexists with His creatures.[94]

Leftow never addresses the theological objection that such an emasculated doctrine of *creatio ex nihilo* does not do justice to the biblical data, which give us clearly to understand that God and the universe do not timelessly co-exist, but that the actual world includes a state of affairs which is God's existing alone without the universe. Typically such a state is described in the ordinary language of the biblical authors as obtaining "before" the world began (John 17:24; Eph. 1:4; 1 Pet. 1:20; cf. Matt. 13:35; 24:21; 25:34; Luke 11:50; Heb. 9:26; Rev. 13:8; 17:8). To quote again the Psalmist's words: "Before the mountains were brought forth, or ever thou hadst formed the earth and the world, from everlasting to everlasting, thou art God" (Ps. 90:2, KJV). Jude's doxology is especially interesting: "to the only God . . . be glory, majesty, dominion, and authority, *before all time* and *now* and *for ever*" (Jude 25, emphasis added). How these ordinary language expressions are to be formulated philosophically—the Bible is not, as Paul Helm reminds us, a philosophy book from which a doctrine of divine eternity may simply be read off the surface—will be addressed in chapter 6; but their intent is clear and

---

[94] Brian Leftow, *Time and Eternity*, Cornell Studies in Philosophy of Religion (Ithaca, N.Y.: Cornell University Press, 1991), 290-291, 310, cf. 322, where he affirms that God is eternally incarnate in Christ. Cf. also 239, where he affirms that in eternity events are "frozen" in an array of tenseless temporal positions. See also Yates's chapter on timeless creation in John C. Yates, *The Timelessness of God* (Lanham, Md.: University Press of America, 1990), 131-163.

they must be taken seriously.[95] The notion that God and the universe timelessly co-exist in an asymmetrical relation of ontological dependence is not only foreign to but actually incompatible with the biblical writers' conception of *creatio ex nihilo,* of God's existing alone and bringing the world into being out of nothing.

Not only so, but the idea that God and creation tenselessly co-exist seems to negate God's triumph over evil. On the static theory of time, evil is never really vanquished from the world: It exists just as sturdily as ever at its various locations in space-time, even if those locations are all earlier than some point in cosmic time (for example, Judgment Day). Creation is never really purged of evil on this view; at most it can be said that evil only infects those parts of creation which are earlier than certain other events. But the stain is indelible. What this implies for events such as the crucifixion and resurrection of Christ is very troubling. In a sense Christ hangs permanently on the cross, for the dreadful events of A.D. 30 never fade away or transpire. The victory of the resurrection becomes a hollow triumph, for the spatio-temporal parts of Jesus that were crucified and buried remain dying and dead and are never raised to new life. It is unclear how we can say with Paul, "Death is swallowed up in victory!" (1 Cor. 15:54) when, on a static theory of time, death is never really done away with.

A robust doctrine of creation, then, involves more than just the tenseless ontological dependence of the world on God. It involves the affirmation that God brings it about that the world comes into being at some time t. Something comes into being at a time t if and only if the following three conditions are met: (i) The thing exists at t, (ii) t is the first time at which the thing exists, and (iii) the thing's existing at t is a tensed fact.[96] The static time theorist cannot affirm that the world came into being at the first moment of its existence and therefore cannot affirm that God created the world in the full sense of the word "create." It seems to me, therefore, that a static conception of time is theologically unacceptable. A robust doctrine of creation requires a dynamic theory of time.

In conclusion, the static conception of time has little to commend it, being based primarily on a Minkowskian, space-time interpretation of Relativity

---

[95] Consideration of *creatio ex nihilo* raises a nest of intriguing and difficult questions: Did God exist in time before the creation of the universe? Does *creatio ex nihilo* imply the creation of time itself? Can God's priority to time be understood in some way other than chronological? I shall try to address these questions in the next chapter. For now it is enough to realize that the biblical writers' expressions about God's existing and planning "before" creation clearly mean to affirm that in some sense God was alone and then brought the world into being out of nothing.

[96] If God can re-create things, then we would have to add to condition (ii) "or t is preceded by a time at which the thing did not exist."

Theory, an interpretation which we are under no constraint to adopt. On the other hand, the static conception of time faces philosophical and theological difficulties that are truly formidable: It "spatializes" time; it gives an incoherent account of the experience of becoming; its analysis of intrinsic change implies the bizarre and multiply flawed doctrine of perdurantism; and it emasculates the biblical doctrine of *creatio ex nihilo*.

Weighing the arguments for and against the dynamic or static conception of time respectively, we seem to have good grounds for believing what people have intuitively always believed: that time is tensed and temporal becoming is real. The dynamic conception of time is correct.

Recommended Reading

INTERPRETATIONS OF RELATIVITY THEORY

Balashov, Yuri. "Enduring and Perduring Objects in Minkowski Space-Time." *Philosophical Studies* 99 (2000): 129-166.

Builder, Geoffrey. "Ether and Relativity." *Australian Journal of Physics* 11 (1958): 279-297, reprinted in *Speculations in Science and Technology* 2 (1979): 230-242.

————. "The Constancy of the Velocity of Light." *Australian Journal of Physics* 11 (1958): 457-480, reprinted in *Speculations in Science and Technology* 2 (1971): 421-437.

Nerlich, Graham. *What Spacetime Explains*. Cambridge: Cambridge University Press, 1994.

————. "Time as Spacetime." In *Questions of Time and Tense*, pp. 119-134. Edited by Robin Le Poidevin. Oxford: Clarendon, 1998.

Prokhovnik, Simon J. *Light in Einstein's Universe*. Dordrecht, Holland: D. Reidel, 1985.

Taylor, Edwin F., and John Archibald Wheeler. *Spacetime Physics*. San Francisco: W. H. Freeman, 1966.

THE MIND-DEPENDENCE OF BECOMING

Black, Max. Review of *The Natural Philosophy of Time*, by G. J. Whitrow. *Scientific American* 206 (April 1962), pp. 179-184.

Cleugh, Mary F. *Time and Its Importance in Modern Thought*. With a foreword by L. Susan Stebbing. London: Methuen, 1937.

Kroes, Peter. *Time: Its Structure and Role in Physical Theories*. Synthese Library 179. Dordrecht: D. Reidel, 1985.

McGilvray, James A. "A Defense of Physical Becoming." *Erkenntnis* 14 (1979): 275-299.

Geach, Peter. "Some Problems about Time." In *Logic Matters*, pp. 302-318. Berkeley: University of California Press, 1972.

"Spatializing" Time

Gale, Richard M. *The Language of Time,* pp. 27-28, 86-100. International Library of Philosophy and Scientific Method. London: Routledge & Kegan Paul, 1968.

McTaggart, J. Ellis. "The Unreality of Time." *Mind* 17 (1908): 457-474.

Mellor, D. H. *Real Time.* Cambridge: Cambridge University Press, 1981.

Craig, William Lane. "Tense and Temporal Relations." *American Philosophical Quarterly* (forthcoming).

The Problem of Intrinsic Change

Carter, William R. and Scott Hestevold. "On Passage and Persistence." *American Philosophical Quarterly* 31 (1994): 269-283.

Hoy, Ronald C. "Becoming and Persons." *Philosophical Studies* 34 (1978): 269-280.

Lewis, David. *On the Plurality of Worlds,* pp. 199-204. Oxford: Basil Blackwell, 1986.

Lewis, Delmas. "Persons, Morality, and Tenselessness." *Philosophy and Phenomenological Research* 47 (1986): 305-309.

Merricks, Trenton. "Endurance and Indiscernibility." *Journal of Philosophy* 91 (1994): 165-184.

Van Inwagen, Peter. "The Doctrine of Arbitrary Undetached Parts." *Pacific Philosophical Quarterly* 62 (1981): 123-137.

_____. "Four-Dimensional Objects." *Noûs* 24 (1990): 245-255.

Creatio ex Nihilo

Westermann, Claus. *Genesis 1–11.* Translated John J. Scullion. Minneapolis: Augsburg, 1984.

Copan, Paul, "Is *Creatio ex nihilo* a Post-biblical Invention?" *Trinity Journal* 17 (1996): 77-93.

Craig, William Lane. "Creation and Conservation Once More." *Religious Studies* 34 (1998): 177-188.

6

GOD, TIME, AND CREATION

OUR STUDY OF GOD and time is nearly complete. We have examined rival conceptions of time, the dynamic versus the static, and have concluded that time is dynamic: Tensed facts and temporal becoming are real. But then it follows from God's creative activity and omniscience that, given the existence of a temporal world, God is also temporal. God quite literally exists now.

Since God never begins to exist nor ever ceases to exist, it follows that God is omnitemporal. He exists at every time that ever exists; that is to say, He endures throughout all eternity. This might seem to imply that God has existed for infinite time in the past and will exist for infinite time in the future. But what if the temporal world has not always existed? According to the Christian doctrine of creation, the world is not infinite in the past but was brought into being out of nothing a finite time ago. Did time itself also have a beginning? Did God exist literally before creation or is He timeless without the world?

I. Did Time Begin?

According to current cosmological theory, time and space came into existence with the Big Bang. The standard model of the expanding universe predicts that in the past the universe was denser than it is today. This has implications for the finitude of past time. As the British physicist P. C. W. Davies explains,

> If we extrapolate this prediction to its extreme, we reach a point when all distances in the universe have shrunk to zero. An initial cosmological singularity therefore forms a past temporal extremity to the universe. We cannot continue physical reasoning or even the concept of spacetime, through such an extremity. For this reason most cosmologists think of the initial singularity as the beginning of the universe. On this view the big bang repre-

sents the creation event, the creation not only of all the matter and energy in the universe, but also of spacetime itself.[1]

On such an understanding, the universe did not spring into being at a point in a previously existing empty space. Rather space and time themselves came into being along with the universe, which implies creation out of absolutely nothing. Thus, Barrow and Tipler assert, "At this singularity, space and time came into existence; literally nothing existed before the singularity, so, if the Universe originated at such a singularity, we would truly have a creation *ex nihilo.*"[2]

This feature of the standard Big Bang model has appeared especially baffling to philosophically minded cosmologists, particularly those with an atheistic bent. For example, the Russian astrophysicist Andrei Linde acknowledges quite frankly the problem the standard model poses for him: "The most difficult aspect of this problem is not the existence of the singularity itself, but the question of what was *before* the singularity.... This problem lies somewhere at the boundary between physics and metaphysics."[3] In order to avoid this question, Linde therefore proposed an eternal inflationary model of the universe, according to which our observable universe was birthed by a prior universe, which was born of a yet prior universe, and so on *ad infinitum.* But in 1994 two other cosmologists, Arvind Borde and Alexander Vilenkin, showed that inflationary scenarios such as Linde's cannot avoid an initial singularity. They conclude, "A physically reasonable spacetime that is eternally inflating to the future must possess an initial singularity.... The fact that inflationary spacetimes are past incomplete forces one to address the question of what, if anything, came before?"[4]

Other cosmologists have tried to eliminate the initial space-time singularity by introducing speculations about quantum gravity, as in Stephen Hawking's famous theory. On such models, imaginary numbers are assigned to the time variable in the equations, which has the effect of suppressing the singular point. But as Hawking himself acknowledges, such models in "imaginary time" are not realistic descriptions of the universe but have mere instrumental value.[5] "When one goes back to the real time in which we live,"

---

[1] P. C. W. Davies, "Spacetime Singularities in Cosmology and Black Hole Evaporations," in *The Study of Time III,* ed. J. T. Fraser, N. Lawrence, and D. Park (Berlin: Springer, 1978), 78-79.
[2] John Barrow and Frank Tipler, *The Anthropic Cosmological Principle* (Oxford: Oxford University Press, 1986), 442.
[3] Andrei Linde, "The Inflationary Universe," *Reports on Progress in Physics* 47 (1984): 976.
[4] Arvind Borde and Alexander Vilenkin, "Eternal Inflation and the Initial Singularity," *Physical Review Letters* 72 (1994): 3305, 3307.
[5] Stephen Hawking and Roger Penrose, *The Nature of Space and Time* (Princeton: Princeton University Press, 1996), 3-4, 121; cf. 53-55.

Hawking admits, "there will still appear to be singularities."[6] In any case, as Barrow emphasizes, such models still involve a merely finite past and so imply the beginning of space and time: "This type of quantum universe has not always existed; it comes into being just as the classical cosmologies could, but it does not start at a Big Bang where physical quantities are infinite. . . ."[7] Thus, the beginning of time is not avoided.

So today, in Hawking's words, "almost everyone now believes that the universe, and *time itself,* had a beginning at the big bang."[8] This consensus seems to lend strong support to the view that neither events nor time existed prior to creation. As physicist David Park says, "It is deceptively easy to imagine events before the big bang . . . , but in physics there is no way to make sense of these imaginings."[9]

The fly in the ointment, however, is Park's phrase "in physics." For we have been wont to emphasize throughout this study that time as it plays a role in physics is at best a measure of time, not time itself. It is perfectly coherent to imagine non-physical events prior to the Big Bang, whether mental events in God's stream of consciousness or events in angelic realms created by God prior to the physical universe. At most, then, the physical evidence proves that physical time had a beginning at the Big Bang, not that time itself so began. In order to explore that question, we shall have to have recourse to metaphysical, rather than physical, arguments.

1. Arguments for the Infinitude of the Past

What reasons, then, might be given for thinking that past time is infinite? We have seen that Newton believed space and time to be infinite because they are the effects of an omnipresent and eternal God.[10] Newton assumed that God cannot exist timelessly and spacelessly. But he gave no arguments at all for this presupposition. We have seen, to the contrary, that there is no reason to think that a personal being could not exist timelessly in the absence of any physical universe.[11] So long as God exists changelessly, He can, in the absence of a temporal world, exist timelessly.

Is there, then, some non-theological reason to think that time is infinite?

---

[6] Stephen Hawking, *A Brief History of Time: From the Big Bang to Black Holes,* with an introduction by Carl Sagan (New York: Bantam Books, 1988), 139.
[7] John Barrow, *Theories of Everything* (Oxford: Clarendon, 1991), 68.
[8] Hawking and Penrose, *Nature of Space and Time,* 20.
[9] David Park, "The Beginning and End of Time in Physical Cosmology," in *The Study of Time IV,* ed. J. T. Fraser, N. Lawrence, and D. Park (Berlin: Springer, 1981), 112-113.
[10] See chapter 2, pages 45-46.
[11] See chapter 3, pages 77-86.

Oxford University's Richard Swinburne thinks so.[12] He argues that at every instant it is true that "There were swans or there were not swans." Necessarily, one of these alternatives is true. But whichever alternative is true, there must be a past at that instant, since the statement is in the past tense. Thus, there cannot be an absolutely first instant of time. Time is unbounded and therefore infinite.

This argument, however, is multiply flawed. In the first place, at best all the argument proves is that every instant of time is preceded by another instant and that therefore there is no first instant of time. But there being no first instant of time is perfectly compatible with the finitude of the past. Compare the series of fractions converging toward zero as a limit: . . . 1/8, 1/4, 1/2. For any fraction you pick, there is always a fraction before it. There is no first fraction in such a series. Nonetheless, the distance covered by all the fractions is still finite. If we let each fraction represent an instant, then we can see that in, say, the first half-minute of time, any instant you pick is preceded by another instant, but the past is not for all that infinite—on the contrary, it is only thirty seconds long. In short, for time to have a beginning, it does not have to have a beginning *point*. Time begins to exist just in case there is some finite interval of time which is not preceded by an interval of equal length. Thus, if time had a beginning, there would be a first hour, or a first minute, or a first second, but there need not be a first instant.

In the second place, the argument fails in any case to show that there cannot be a first instant of time. Swinburne argues that the truth of past-tense statements requires that there be a past. But specialists in the logic of tensed sentences have shown that this is not correct. A statement such as "There were swans" can be analyzed logically as asserting, "It was the case that there are swans." Such a statement does imply a past. But a negative statement such as "There were no swans," when asserted at the first instant of time, should not be analyzed as asserting, "It was the case that there are no swans," but rather as asserting, "It was not the case that there are swans." Such a statement does not imply a past, for it denies that it was the case that swans exist. Hence, such a statement can be true at a first instant of time.

Thus, we have not seen any good reasons for thinking that the past is or must be infinite.

2. *Arguments for the Finitude of the Past*

Are there any good reasons for thinking that the past is finite? There is a long tradition in Western philosophy of arguments for the finitude of the past, and

---

[12] Richard Swinburne, *Space and Time,* 2d ed. (New York: St. Martin's Press, 1981), 172.

while most philosophers today are skeptical of such proofs, it does seem to me that some of these arguments, at least, are quite plausible and that the standard refutations of them fail.[13]

The Impossibility of an Actual Infinite

Consider, for example, the argument based on *the impossibility of the existence of an actual infinite*. This argument may be formulated as follows:

1. An actual infinite cannot exist.

2. A beginningless series of equal past intervals of time is an actual infinite.

3. Therefore, a beginningless series of equal past intervals of time cannot exist.

But if there cannot be a beginningless series of equal past intervals of time, then time must have begun to exist.

In order to understand this argument, let us examine each step individually.

1. *An actually infinite number of things cannot exist.* In order to understand this first step, we need to understand what an actual infinite is. There is a difference between a potential infinite and an actual infinite. A potential infinite is a collection that is increasing toward infinity as a limit but never gets there. Such a collection is really indefinite, not infinite. For example, any finite distance can be subdivided into potentially infinitely many parts. You can just keep on dividing parts in half forever, but you will never arrive at an actual "infinitieth" division or come up with an actually infinite number of parts. By contrast, an actual infinite is a collection in which the number of members really is infinite. The collection is not growing toward infinity; it *is* infinite, it is "complete." This sort of infinity is used in set theory to designate sets that have an infinite number of members, such as {1, 2, 3 . . . }. Now the argument is, not that a potentially infinite number of things cannot exist, but that an actually infinite number of things cannot exist. For if an actually infinite number of things could exist, this would spawn all sorts of absurdities.

Perhaps the best way to bring this home is by means of an illustration.

---

[13] See William Lane Craig, *The Kalam Cosmological Argument,* Library of Philosophy and Religion (London: Macmillan, 1979); William Lane Craig and Quentin Smith, *Theism, Atheism, and Big Bang Cosmology* (Oxford: Clarendon, 1993).

Let me use one of my favorites, Hilbert's Hotel, a product of the mind of the great German mathematician David Hilbert.[14] Let us first imagine a hotel with a finite number of rooms. Suppose, furthermore, that all the rooms are full. When a new guest arrives asking for a room, the proprietor apologizes, "Sorry, all the rooms are full," and the new guest is turned away. But now let us imagine a hotel with an infinite number of rooms and suppose once more that *all the rooms are full.* There is not a single vacant room throughout the entire infinite hotel. Now suppose a new guest shows up, asking for a room. "But of course!" says the proprietor, and he immediately shifts the person in room #1 into room #2, the person in room #2 into room #3, the person in room #3 into room #4, and so on, out to infinity. As a result of these room changes, room #1 now becomes vacant and the new guest gratefully checks in. But remember: Before he arrived, all the rooms were full!

Equally curious, according to the mathematicians, there are now no more persons in the hotel than there were before: the number is just infinite. But how can this be? The proprietor just added the new guest's name to the register and gave him his keys—how can there not be one more person in the hotel than before?

But the situation becomes even stranger. For suppose an infinity of new guests show up at the desk, asking for rooms. "Of course, of course!" says the proprietor, and he proceeds to shift the person in room #1 into room #2, the person in room #2 into room #4, the person in room #3 into room #6, and so on out to infinity, always putting each former occupant into the room number twice his own. Because any natural number multiplied by two always equals an even number, all the guests wind up in even-numbered rooms. As a result, all the odd-numbered rooms become vacant, and the infinity of new guests is easily accommodated. And yet, before they came, all the rooms were full! And again, strangely enough, the number of guests in the hotel is the same after the infinity of new guests check in as before, even though there were as many new guests as old guests. In fact, the proprietor could repeat this process *infinitely many times* and yet there would never be one single person more in the hotel than before.

But Hilbert's Hotel is even stranger than the German mathematician made it out to be. For suppose some of the guests start to check out. Suppose the guest in room #1 departs. Is there not now one less person in the hotel? Not according to the mathematicians—but just ask Housekeeping! Suppose

---

[14] The illustration of Hilbert's Hotel is related in George Gamow, *One, Two, Three, Infinity* (London: Macmillan, 1946), 17.

the guests in rooms #1, 3, 5 . . . check out. In this case an infinite number of people have left the hotel, but according to the mathematicians, there are no less people in the hotel—but don't talk to Housekeeping! In fact, we could have every other guest check out of the hotel and repeat this process infinitely many times, and yet there would never be any fewer people in the hotel.

Now suppose the proprietor doesn't like having a half-empty hotel (it looks bad for business). No matter! By shifting occupants as before, but in reverse order, he transforms his half-vacant hotel into one that is jammed to the gills. You might think that by these maneuvers the proprietor could always keep this strange hotel fully occupied. But you would be wrong. For suppose that the persons in rooms #4, 5, 6 . . . checked out. At a single stroke the hotel would be virtually emptied, the guest register would be reduced to three names, and the infinite would be converted to finitude. And yet it would remain true that the *same* number of guests checked out this time as when the guests in rooms #1, 3, 5 . . . checked out! Can anyone believe that such a hotel could exist in reality?

Hilbert's Hotel is absurd. As one person remarked, if Hilbert's Hotel could exist, it would have to have a sign posted outside: NO VACANCY—GUESTS WELCOME. The above sorts of absurdities show that it is impossible for an actually infinite number of things to exist. There is simply no way to avoid these absurdities once we admit the possibility of the existence of an actual infinite. Students sometimes react to such absurdities as Hilbert's Hotel by saying that we really don't understand the nature of infinity and, hence, these absurdities result. But this attitude is simply mistaken. Infinite set theory is a highly developed and well-understood branch of mathematics, so that these absurdities result precisely because we *do* understand the notion of a collection with an actually infinite number of members.

Critics have raised various objections to premise (1). For example, the philosopher Wallace Matson objects that (1) must mean that an actual infinite is *logically* impossible; but it is easy to show that such a collection is logically possible. For example, the set of negative numbers { . . . , -3, -2, -1} is an actually infinite collection with no first member.[15] Similarly, the Australian philosopher Graham Oppy insists that because infinite set theory is a logically consistent system, it must be possible for an actual infinite to exist.[16]

---

[15] Wallace Matson, *The Existence of God* (Ithaca, N.Y.: Cornell University Press, 1965), 58-60. For discussion see William Lane Craig, "Wallace Matson and the Crude Cosmological Argument," *Australasian Journal of Philosophy* 57 (1979): 167-170.

[16] Graham Oppy, "Craig, Mackie, and the *Kalam* Cosmological Argument," *Religious Studies* 27 (1991): 193-195. For discussion see William Lane Craig, "Graham Oppy on the Kalam Cosmological Argument," *Sophia* 32 (1993): 1-11.

The error of these thinkers lies in their failure to distinguish between what philosophers have called "strict logical possibility" and "broad logical possibility." Something is strictly logically possible if it does not involve a contradiction. Something may be strictly logically possible, however, without being capable of existing in reality. For example, there is no logical contradiction in asserting, "Something has a shape but not a size," "An event occurs before itself," or, "Something came into existence without a cause," but all of these statements are plausibly broadly logically impossible. Broad logical possibility is thus usually identified with metaphysical possibility, that is to say, with what is possible in reality. Now infinite set theory is strictly logically consistent, granted its axioms and rules, but that does nothing to prove that such a system can exist in the real world. This fact is especially evident when it comes to mathematical operations such as subtraction and division, which transfinite arithmetic must prohibit in order to maintain logical consistency. While we can slap the hand of the mathematician who attempts such operations with infinite numbers, we cannot in reality prevent people from checking out of a Hilbert's Hotel with all the attendant absurdities.

One should also note that even the mathematical existence of the actual infinite cannot just be taken for granted. For the small, but brilliant, school of intuitionist mathematicians denies even mathematical infinities. In their view the number series is merely potentially infinite, not actually infinite. So long as intuitionism remains a viable position in the philosophy of mathematics, one cannot justifiably appeal to mathematical infinites as counter-examples to premise (1).

Some critics have claimed that even the existence of a potential infinite implies the existence of an actual infinite. For example, Rucker states that in order for the intuitionist to regard the number series as potentially infinite via the operation of counting, there must exist a "definite class of possibilities" which is actually infinite.[17] Similarly, Sorabji thinks that a line's being potentially infinitely divisible implies that there is an actually infinite number of positions where the line could be divided.[18]

But these inferences are mistaken. Infinite divisibility does not imply an infinite number of pre-existing points unless one presupposes that a line is already composed of an infinite number of points. But if an extension is logically prior to any points one specifies in it, potential infinite divisibility does

---

[17] Rudolf v. B. Rucker, "The Actual Infinite," *Speculations in Science and Technology* 3 (1980): 66. For a discussion see William Lane Craig, "Time and Infinity," *International Philosophical Quarterly* 31 (1991): 387-401.
[18] Richard Sorabji, *Time, Creation, and the Continuum* (Ithaca, N.Y.: Cornell University Press, 1983), 210-213, 322-324.

not imply the existence of points. From the truth that "Possibly, there is some point at which a line is divided," it does not follow that "There is some point at which the line is possibly divided." The same is the case for numbers and counting.[19]

The late Oxford University philosopher J. L. Mackie disputed premise (1) because the so-called absurdities of the existence of an actual infinite are resolved once we understand that for infinite groups Euclid's axiom "The whole is greater than its part" does not hold as it does for finite groups.[20] Similarly, Quentin Smith comments that once we understand that an infinite set has a proper subset which has the same number of members as the set itself, then the purportedly absurd situations become "perfectly believable."[21] But far from being the solution, this is precisely the problem. Because in infinite set theory this axiom is denied, one winds up with all sorts of absurdities such as Hilbert's Hotel when one tries to translate that theory into reality. The issue is not whether these consequences would result if an actual infinite were to exist; we agree that they would. The question is whether such consequences are metaphysically possible. That question is not resolved by reiterating that they would be possible if an actual infinite could exist. Moreover, not all the absurdities result from a denial of Euclid's axiom: The absurdities illustrated by guests checking out of Hilbert's Hotel result from subtraction of infinite quantities, which set theory must prohibit to maintain logical consistency.

Hilbert himself, who declared that Cantor's infinite set theory is "one of the supreme achievements of purely intellectual human activity," and that "No one shall drive us out of the paradise which Cantor has created for us,"[22] nevertheless also believed that that paradise exists only in the realm of the intellect: "the infinite is nowhere to be found in reality. It neither exists in nature nor provides a legitimate basis for rational thought. . . . The role that remains for the infinite to play is solely that of an idea. . . ."[23] The great philosopher Ludwig Wittgenstein agreed—though with considerably less enthusiasm for Cantor's paradise. Reacting to Hilbert's remark about the Cantorian paradise, Wittgenstein quipped, "I would say, 'I wouldn't dream

---

[19] The deeper issue here is whether abstract entities such as points, numbers, and sets actually exist in a mind-independent way. For a brief discussion, see William Lane Craig, "A Swift and Simple Refutation of the *Kalam* Cosmological Argument?" *Religious Studies* 35 (1999): 57-72.

[20] J. L. Mackie, *The Miracle of Theism* (Oxford: Clarendon, 1982), 93. For discussion see William Lane Craig, "Prof. Mackie and the *Kalam* Cosmological Argument," *Religious Studies* 20 (1985): 367-375.

[21] Quentin Smith, "Infinity and the Past," *Philosophy of Science* 54 (1987): 69.

[22] David Hilbert, "On the Infinite," in *Philosophy of Mathematics*, ed. with an introduction by Paul Benacerraf and Hilary Putnam (Englewood Cliffs, N.J.: Prentice-Hall, 1964), 139, 141.

[23] Ibid., 151.

of trying to drive anyone from this paradise.' I would do something quite different: I would try to show you that it is not a paradise—so that you'll leave of your own accord. I would say, 'You're welcome to this; just look about you.'"[24] Once we take a sober look about us at the absurd consequences of such a world, we shall not regret leaving it to exist only in the imagination.

Premise (2) seems pretty obvious. If time never had a beginning, then if one were to add up all the temporal intervals of some finite extent, say, seconds, then there will have existed an actually infinite number of seconds prior to the present second.

As I say, this seems pretty obvious. Nevertheless, some critics of the argument have denied premise (2), claiming that the past is a potential infinite only. Swinburne, for example, admits that it makes little sense to think that the past could have an end but no beginning, but he advises that we avoid this puzzle by numbering the events of the past by beginning in the present and proceeding to count backwards in time.[25] In this way the past is converted from a series with no beginning but an end, into a series with a beginning but no end, which is unobjectionable.

It seems to me that this solution is clearly wrong-headed. In order for the past to be a mere potential infinite, it would have to be finite, but growing in a backwards direction. This contradicts the nature of time and becoming. Swinburne confuses the mental regress of counting with the real progress of time itself. The direction of time itself is from past to future, so that if the series of past seconds is beginningless, then an actually infinite number of seconds have elapsed.

Thus it seems to me that the objections lodged against the argument's premises are less plausible than those premises themselves. Given the truth of the two premises, it follows that a beginningless series of equal past intervals of time cannot exist. Thus, time must have had a beginning.

THE IMPOSSIBILITY OF THE FORMATION OF AN ACTUAL INFINITE

But suppose the above is altogether wrong. Suppose that an actual infinite can exist. That still does not imply that the past can be actually infinite. For we now must consider how an actual infinite can come to exist. And here we confront the argument for the finitude of the past based on *the impossibility of the formation of an actual infinite by successive addition*. This argument goes as follows:

---

[24] Ludwig Wittgenstein, *Lectures on the Foundations of Mathematics*, ed. Cora Diamond (Sussex: Harvester Press, 1976), 103.
[25] Swinburne, *Space and Time*, 298-299.

1. A collection formed by successive addition cannot be actually infinite.

2. The series of equal past intervals of time is a collection formed by successive addition.

3. Therefore, the series of equal past intervals of time cannot be actually infinite.

Once again, if the series of equal, past temporal intervals is not actually infinite, then the past must be finite.

Premise (1) is the crucial step in this argument. One cannot form an actually infinite collection of things by successively adding one member after another. Since one can always add one more before arriving at infinity, it is impossible to reach actual infinity. Sometimes this is called the impossibility of "counting to infinity" or "traversing the infinite." It is important to understand that this impossibility has nothing to do with the amount of time available: It belongs to the nature of infinity that it cannot be so formed.

Now someone might say that while an infinite collection cannot be formed by beginning at a point and adding members, nevertheless an infinite collection could be formed by never beginning but ending at a point, that is to say, ending at a point after having added one member after another from eternity. But this method seems even more unbelievable than the first method. If one cannot count *to* infinity, how can one count down *from* infinity? If one cannot traverse the infinite by moving in one direction, how can one traverse it simply by moving in the opposite direction?

Indeed, the idea of a beginningless series ending in the present seems absurd. To give just one illustration: Suppose we meet a man who claims to have been counting from eternity and is now finishing: . . . , -3, -2, -1, 0. We could ask, why did he not finish counting yesterday or the day before or the year before? By then an infinite time had already elapsed, so that he should already have finished by then. Indeed, at no point in the infinite past could we ever find the man finishing his countdown, for at any point he would already be done! In fact, no matter how far back into the past we go, we can never find the man counting at all, for at any point we reach he will have already finished. But if at no point in the past do we find him counting, this contradicts the hypothesis that he has been counting from eternity. This illustrates the fact that the formation of an actual infinite by successive addition is equally impossible whether one proceeds to or from infinity.

Sorabji has objected that, while it is true that at any point in the past someone counting down from infinity will have counted an infinity of negative numbers, there is no reason to think that he will have counted *all* the negative numbers.[26] David Conway also claims that there is no good reason to think that if someone had counted an infinite number of numbers by yesterday, then he would have finished his countdown of all the numbers by yesterday.[27] But these objections misunderstand the argument. At whatever number the countdown of the negative numbers ends—say, -17 or -3—an actual infinite will have been completed by successive steps, and we may ask why that task was not accomplished by yesterday. The central contention of the argument is not that if someone had counted an infinite number of numbers by yesterday, then he would have finished counting all the numbers by yesterday, but rather that if someone would have finished his countdown of the negative numbers by today, then he would have already finished that same countdown by yesterday. The defender of the infinite past claims that the seemingly impossible task of counting all the negative numbers is possible because for every negative number to be counted there is a corresponding moment of past time in which to count it. But that is true at every moment of the infinite past! Thus, if the task could be completed today, it becomes inexplicable why it was not already completed yesterday, or the day before yesterday, *ad infinitum*. On Sorabji and Conway's position, it becomes inexplicable why the counter should finish at any number that he does.

Moreover, a deeper absurdity now surfaces: Suppose we have two counters, one man counting a negative number every second and the other counting a negative number every hour. Since the number of past seconds and the number of past hours is identical (if the past is infinite), both men complete their countdown at the same time. But this is absurd, since one man is counting down the numbers 3,600 times faster than the other!

Critics typically allege that premise (1) illicitly assumes an infinitely distant starting point in the past and then pronounces it impossible to travel from that point to today.[28] If we take the notion of infinity "seriously," Mackie says, then we must say that in the infinite past there would be no start-

---

[26] Sorabji, *Time, Creation, and the Continuum*, 219-222. For discussion see William Lane Craig, critical notice of *Time, Creation, and the Continuum*, International Philosophical Quarterly 25 (1985): 319-326.
[27] David A. Conway, "'It Would Have Happened Already': On One Argument for a First Cause," *Analysis* 44 (1984): 159-166.
[28] Mackie, *Miracle of Theism*, 93.

ing point whatever, not even an infinitely distant one. Yet from any given point in the past, there is only a finite distance to the present.

But it seems to me that Mackie's allegation that the argument presupposes an infinitely distant starting point in the past is entirely groundless. As I have explained it, the argument concerns the possibility of completing a task such as counting down all the negative numbers in succession, a series which has no beginning. Indeed, the fact that the past has *no beginning at all*, not even an infinitely distant one, makes it even more bewildering how the past could have been formed by successive addition. It is like trying to jump out of a bottomless pit! And Mackie's observation that from any point in the past the distance to the present is finite is quite correct but simply irrelevant to the discussion. For the issue is how the *whole* infinite past can be formed by successive addition, not merely some finite portion of it. Does Mackie think that because every *finite* segment of the past can be formed by successive addition, the whole infinite past can be formed by successive addition? That is as logically fallacious as saying that because every part of an elephant is light in weight, therefore the whole elephant is light in weight.

It seems to me, therefore, that premise (1), despite the objections of its detractors, is more plausible than its denial.

As for premise (2), the only persons who deny this step of the argument are the proponents of a static conception of time. Since they reject the reality of temporal becoming, they deny that the past was formed by successive addition. All times exist tenselessly, and there is no lapse of time. But our lengthy inquiry into the nature of time in chapters 4 and 5 brought us to the conclusion that the static conception of time is wrong. Time is dynamic, and therefore the past has been formed sequentially, one moment elapsing after another. If the past is infinite, then God has lived through an infinite number of past temporal intervals one at a time in order to arrive at today. But such a traversal of the infinite past, as we have seen, seems absurd.

From the two premises of the argument it again follows that the series of equal past intervals of time cannot be actually infinite.

We thus have what seem to me to be two quite plausible, independent arguments for the finitude of time. The objections of the argument's critics do not strike me as compelling, so that the premises of the argument remain more plausible than their denials.

Why Did God Not Create the World Sooner?

Moreover, there is a third, peculiarly theological argument for the finitude of past time which bedevils proponents of Newtonian eternity, namely, why did God not create the world sooner? Leibniz pressed this question in his famous correspondence with Newton's follower Samuel Clarke.[29] On Leibniz's relational view of time, time does not exist in the absence of events. Hence, time begins at the moment of creation, and it is simply maladroit to ask why God did not create the world sooner, since there is no "sooner" prior to the moment of creation. Time comes into existence with the universe, and so it makes no sense to ask why it did not come into being at an earlier moment. But on Newton's view, God has endured through an infinite period of creative idleness up until the moment of creation. Why did He wait so long?

This problem can be formulated as follows (letting t represent any time prior to creation and n represent some finite amount of time):

1. If the past is infinite, then at t God delayed creating until $t + n$.

2. If at t God delayed creating until $t + n$, then He must have had a good reason for doing so.

3. If the past is infinite, God cannot have had a good reason for delaying at t creating until $t + n$.

4. Therefore, if the past is infinite, God must have had a good reason for delaying at t, and God cannot have had a good reason for delaying at t.

5. Therefore, the past is not infinite.

Premise (1) is obviously true, given a dynamic view of time. At t God could have created the world. But He did not. He deliberately waited until a later time. He self-consciously refrained from creating at t and delayed His action until $t + n$.

Premise (2) seems to be plausibly true. It does not depend for its truth on the validity of some broader Principle of Sufficient Reason (a controversial principle defended by Leibniz to the effect that everything has a reason why it is the way it is). Rather it states merely that in this specific case, God, in deciding to delay creating the world until some later time,

---

[29] G. W. Leibniz, "Mr. Leibniz's Third Paper," in *The Leibniz-Clarke Correspondence*, ed. with an introduction and notes by H. G. Alexander (Manchester: Manchester University Press, 1956), 42.

must have had some good reason for doing so. Notice also that premise (2) does not presuppose either the finitude or the infinitude of the past. It merely asserts that if at some moment prior to creation God deliberately deferred creating until a later moment, then He must have had a reason for doing so. A perfectly rational person does not delay some action which he wills to undertake unless he has a good reason for doing so. Since God is a supremely rational being, premise (2) strikes me as eminently plausible.

That brings us to premise (3), which again seems obviously true. As Leftow points out in his interesting analysis of this problem,[30] if God comes to acquire at some moment a reason to create the world, this reason must be due to some change either in God or in the world. The only change going on outside God is the lapse of absolute time itself. But since all moments of time are alike, there is nothing special about the moment of creation which would cause God to delay creating at t until $t + n$ had arrived. After all, at t God has already waited for infinite time to create the world, so why wait any longer? There is nothing about $t + n$ that makes it a more appropriate time to create than t. As for God Himself, He has from time immemorial been perfectly good, omniscient, and omnipotent, so that there seems to be no change that could occur in Him that would prompt Him to create at some time rather than earlier. Thus, it seems impossible that God should acquire some reason to create which He did not always have; nor by the same token does it seem possible for Him to have always had a reason for singling out $t + n$ as the moment at which to create.

Leftow tries to escape this reasoning by suggesting that God's reason for delaying creation is the joy of anticipation of creating. Just as we find joy in the anticipation of some great good, so God can enjoy the anticipation of creation.

But why would God delay creating for *infinite* time? Having already at t anticipated for infinite time His creating the world, why would He yet delay even longer until $t + n$? Why did He quit anticipating at $t + n$ instead of earlier or later? Leftow answers that there comes a point at which the joy of anticipating begins to fade. So God will not want to delay creating beyond that point. He knows from all eternity precisely when His anticipation peaks and so will not delay beyond that point. Leftow imagines a sort of pleasure curve charting God's rising and falling anticipation of creation (Fig. 6.1).

---

[30] Brian Leftow, "Why Didn't God Create the World Sooner?" *Religious Studies* 27 (1991): 157-172.

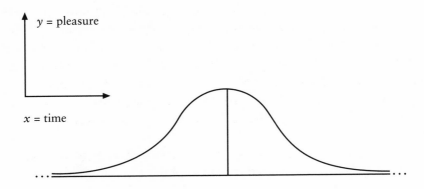

Fig. 6.1: God's anticipatory pleasure rises from a low at $t = -\infty$ to a peak value before declining toward a low at $t = +\infty$.

God will create the world at the moment when His anticipatory pleasure peaks.

We might be skeptical of such a portrayal of God's anticipatory pleasure as grossly anthropomorphic; but never mind. The more serious problem is that Leftow's scheme does not explain anything. For we may still ask, "Why does God's anticipatory pleasure peak at $t + n$ instead of at t?" Leftow answers that since the curve displayed in Fig. 6.1 extends infinitely into the past and future, it cannot be "shifted" in either direction. It is fixed in time and so must peak when it does.

This answer fails to appreciate the paradoxical nature of the actual infinite. Just as Hilbert's Hotel can accommodate infinitely many new guests simply by shifting each guest into a room with a number twice his own, so God's pleasure curve, though infinitely extended, can be shifted backward in time simply by dividing every value of the x-coordinate by two. Since the past is supposed to be actually infinite, there is no danger of "scrunching up" the earlier slope of the curve by such a backward shift. If such a shift seems impossible, this only calls into question once again the infinity of the past. But if the past is infinite, there is no problem. Therefore, Leftow has not provided a good reason for why God at t should delay creating until $t + n$. After all, by t God has already had eternity to anticipate creating the world.

But if premises (1)-(3) are true, then (4) and (5) logically follow. Accordingly, it seems to me that Leibniz was quite correct in opposing Newton with respect to the infinity of God's past.

We thus seem to have three good arguments for denying the infinity of the past and holding to the beginning of time: the impossibility of the existence of an actual infinite; the impossibility of the formation of an actual infinite by successive addition; and the impossibility of God's delaying creation from eternity. For all these reasons it is plausible to believe that time began to exist and that, therefore, God has not existed for infinite time.

II. God and the Beginning of Time

But now we are confronted with an extremely bizarre situation. God exists in time. Time had a beginning. God did not have a beginning. How can these three statements be reconciled? If time began to exist—say, for simplicity's sake, at the Big Bang—then in some difficult-to-articulate sense God must exist *beyond* the Big Bang, alone without the universe. He must be changeless in such a state; otherwise time would exist. And yet this state, strictly speaking, cannot exist before the Big Bang in a temporal sense, since time had a beginning. God must be causally, but not temporally, prior to the Big Bang. With the creation of the universe, time began, and God entered into time at the moment of creation in virtue of His real relations with the created order. It follows that God must therefore be timeless without the universe and temporal with the universe.

Now this conclusion is startling and not a little odd. For on such a view, there seem to be two phases of God's life, a timeless phase and a temporal phase, and the timeless phase seems to have existed earlier than the temporal phase. But this is logically incoherent, since to stand in a relation of *earlier than* is by all accounts to be temporal.[31]

1. Amorphous Time

How are we to escape this apparent antinomy? One possibility is suggested by a re-examination of the three arguments we presented for the finitude of the past. Strictly speaking none of those arguments reached the conclusion, "Therefore, time began to exist." Rather what they proved is that there cannot have been an infinite number of equal temporal intervals in the past. But if we can conceive of a time which is not divisible into intervals, a sort of undifferentiated time, then the arguments are compatible with the existence of that sort of time prior to creation. God existing alone without the universe would exist in an amorphous time before the beginning of divisible time as we know it. A number of philosophers associated with Oxford University

---

[31] See Leftow's statement of the objection in *A Companion to Philosophy of Religion,* ed. Philip L. Quinn and Charles Taliaferro, Blackwell's Companions to Philosophy 8 (Oxford: Blackwell, 1997), s.v. "Eternity," by Brian Leftow.

have defended such a view of divine eternity, so that one could aptly speak of the Oxford school on this issue.[32]

Members of the Oxford school tend to embrace the doctrine of *metric conventionalism* with respect to time. The metric of time has to do with the measure of time—whether two separate intervals of time can be said to be equal or unequal in extent. Conventionalism is the view that there is no objective fact about this matter. There is no objective fact that two separate temporal intervals are equal; it all depends on what measures we use. In the absence of any measures, there is no objective fact that one interval is longer or shorter than another distinct interval. Thus, prior to creation it is impossible to differentiate between a tenth of a second and ten trillion years. There is no moment one hour, say, before creation. Time literally lacks any intrinsic metric.

Such an understanding of God's time prior to creation seems quite attractive. It enables us to speak literally of God's existing before creation. And yet we seem to avoid the problematic claim that God has endured through infinite time prior to creating the universe.

Nevertheless, a closer inspection of the view reveals difficulties. Metric conventionalism is the view that there is no objective fact of the matter concerning the comparative lengths of separate temporal intervals. But metric conventionalism does not hold that there really are no intervals of time or that no intervals can be objectively compared with respect to length. Thus, even in a metrically amorphous time, there are objective factual differences of length for certain temporal intervals (Fig. 6.2).

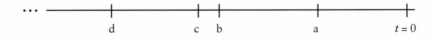

Fig. 6.2: Intervals in a metrically amorphous time prior to the moment of creation $t=0$.

According to metric conventionalism, there is no fact of the matter concerning the comparative lengths dc and cb or db and ca. But there is an objective difference in length between da and ca or cb and ca, namely, $da > ca$ and $cb < ca$. For in the case of intervals which are proper parts of other intervals, the proper parts are factually shorter than their encompassing parts.

But this implies that prior to creation God has endured through a begin-

---

[32] John Lucas, *A Treatise on Time and Space* (London: Methuen, 1973), 311-312; Alan G. Padgett, *God, Eternity, and the Nature of Time* (New York: St. Martin's, 1992), 122-146; Richard Swinburne, "God and Time," in *Reasoned Faith*, ed. Eleonore Stump (Ithaca, N.Y.: Cornell University Press, 1993), 204-222.

ningless series of longer and longer intervals. In fact we can even say that such a time must be infinite. For the past is finite if and only if there is a first interval of time. (An interval is first if there exists no interval earlier than it, or if there exists no interval greater than it but having the same end point.) The metrically amorphous past is clearly not finite. But is it infinite? The past is infinite if and only if there is no first interval of time and time is not circular. Thus, the amorphous time prior to creation would be infinite, even though we cannot compare the lengths of separate intervals within it. Thus, all the difficulties of an infinite past return to haunt us.

The shortcoming of the Oxford school is that it has not been radical enough. It proposes to dispense with the metric of time while still retaining the geometry of time as a line. Since on a geometrical line intervals can be objectively distinguished and, when included in one another, compared in length, time is not sufficiently undifferentiated to avoid the problems attending an infinite past. What must be done is to dissolve the geometrical structure of time as a line. One must maintain that prior to creation there literally are no intervals of time at all. In such a time, there would be no earlier and later, no enduring through successive intervals and, hence, no waiting, no temporal becoming, nothing but the eternal "now." This state would pass away as a whole, not successively, at the moment of creation, when time begins. It would be an undifferentiated "before," followed by a differentiated "after."

The problem is that such a changeless, undifferentiated state looks suspiciously like a state of timelessness! It seems to have the topology of a point, the traditional representation of timeless eternity. The only sense in which such a state can be said to be temporal is that it exists literally before God's creation of the world and the beginning of differentiated time.

2. Timelessness without Creation

Perhaps this realization ought to prompt us to reconsider the alternative that God is simply timeless without creation and temporal subsequent to creation. Detractors of this position simply assume that if God's life lacks earlier and later parts, then it has no phases. But why could there not be two phases of God's life, one timeless and one temporal, which are not related to each other as earlier and later? Critics have perhaps too quickly assumed that if any phase of God's life is timeless, the whole must be timeless.

We have already seen that a state of undifferentiated time looks very much like timelessness. This impression is reinforced by recalling the dynamic theory of time. On a static theory of time, it is very tempting to picture the two phases of God's life as equally existent, bounded by the moment of creation, the one earlier and the other later. But given a dynamic theory of time, this picture is a misrepresentation. In reality God existing without creation is changelessly alone, and no event disturbs this complete tranquility. There is no before, no after, no temporal passage, no future phase of His life. There is just God.

To claim that time would exist without the universe in virtue of the beginning of the world seems to postulate a sort of backward causation: The occurrence of the first event causes time to exist not only with the event but also before it. But on a tensed theory of time, such retrocausation is metaphysically impossible, for it amounts to something's being caused by nothing, since at the time of the effect the retro-cause in no sense exists.[33]

The impression that God without creation is timeless can be reinforced by a thought experiment. Imagine God existing changelessly alone in a possible world in which He refrains from creation. In such a world, God is reasonably conceived to be timeless. But God, actually existing alone without creation, is no different than He would be in such a possible world, even though in the actual world He becomes temporal by creating. Apart from backward causation, there seems to be nothing that would produce a time prior to the moment of creation.

Perhaps an analogy from physical time will be illuminating. The initial Big Bang singularity is not considered to be part of time, but to constitute a boundary to time. Nevertheless, it is causally connected to the universe. In an analogous way, perhaps we could say that God's timeless eternity is, as it were, a boundary of time which is causally, but not temporally, prior to the origin of the universe.

It seems to me, therefore, that it is not only coherent but also plausible that God existing changelessly alone without creation is timeless and that He enters time at the moment of creation in virtue of His real relation to the temporal universe. The image of God existing idly before creation is just that: a figment of the imagination. Given that time began to exist, the most plausible view of God's relationship to time is that He is timeless without creation and temporal subsequent to creation.

Recommended Reading

THE INFINITUDE OF THE PAST

Craig, William Lane. *The* Kalam *Cosmological Argument*. Library of Philosophy and Religion. London: Macmillan, 1979.

---

[33] For discussion see William Lane Craig, *Divine Foreknowledge and Human Freedom*, Brill's Studies in Intellectual History 19 (Leiden: E. J. Brill, 1991), 150-156.

Craig, William Lane, and Quentin Smith. *Theism, Atheism, and Big Bang Cosmology.* London: Clarendon, 1993.

Sorabji, Richard. *Time, Creation, and the Continuum.* Ithaca, N.Y.: Cornell University Press, 1986.

GOD AND THE BEGINNING OF TIME

Craig, William Lane. "Timelessness and Creation." *Australasian Journal of Philosophy* 74 (1996): 646-656.

Padgett, Alan G. *God, Eternity, and the Nature of Time,* chapter 6. New York: St. Martin's, 1992.

Senor, Thomas. "Divine Temporality and Creation *ex Nihilo.*" *Faith and Philosophy* 10 (1993): 86-92.

Swinburne, Richard. "God and Time." In *Reasoned Faith,* pp. 204-222. Edited by Eleonore Stump. Ithaca, N.Y.: Cornell University Press, 1993.

Craig, William Lane, Paul Helm, Alan Padgett, and Nicholas Wolterstorff. *God and Time: Four Views.* Edited by Gregory Ganssle. Downer's Grove, Ill.: InterVarsity Press, 2001.

7

Conclusion

WE HAVE REACHED the end of our long and arduous trail, and now it is time to summarize our argument. Divine eternity is one of those attributes of God which is underdetermined with respect to the biblical data. A literal reading of the biblical texts gives the overriding impression that God is eternal in the sense of existing at all times, not in the sense of being timeless. But there are passages which point in another direction, especially those suggesting that time had a beginning. More importantly, the fact that the biblical authors did not write as philosophers should make us wary of pressing their descriptions of God into categories which may not have been germane to their purposes. The Christian who wants to understand more profoundly the nature of divine eternity and God's relationship to time has no recourse but to reflect philosophically on these issues if he is to come to some well-founded views on such questions.

In our first chapter we examined the principal arguments for God's being timeless and found most of them to be either unsound or inconclusive. Although divine timelessness could be deduced from divine simplicity or immutability, those doctrines are too controversial to serve as a firm foundation for taking divine eternity as timelessness. Although some proponents of divine timelessness have appealed to Relativity Theory in support of their doctrine, that theory can be interpreted along the lines advocated by H. A. Lorentz as a theory about the behavior of clocks and rods in motion, in which case it is entirely compatible with the existence of a privileged, divine time such as Isaac Newton believed in. The one argument for divine timelessness which did have some bite is the argument based on the inherent deficiency of temporal life, whose fleeting nature seems incompatible with the life of a most perfect being. A theorist of tenseless time who holds that God is in time could avert this argument, since he denies the reality of temporal becoming—unless the argument is understood experientially. For even on a tenseless theory of time, a temporal deity will still *experience* the fleeting nature of His life as lived even though none of it actually passes away or comes to be. If the argument is construed experientially, however, then it is not so obvious that an omniscient God, who could bring to mind past and future experiences with a vividness comparable to that of present experiences, should find temporal passage so melancholy an affair. We concluded that the argument from the incompleteness of temporal life might justifiably motivate a doctrine of divine timelessness if there were not off-setting arguments for divine temporality.

The conclusion of this study is that there are, indeed, such arguments. Not that the very idea of a timeless God is incoherent—we examined and dismissed arguments that timelessness and personhood are logically incompatible. Critics of timeless personhood have failed to show either that in order to be personal God has to possess properties inconsistent with timelessness, or that a timeless God cannot possess those properties which are essential to personhood. On the contrary, we saw that it is quite plausible that a timeless being could exemplify properties sufficient for personhood. So it is not true that a timeless God cannot be personal.

Rather I argued that, given the truth of a dynamic or tensed theory of time, God cannot be timeless if a temporal world exists. For if a tensed theory of time is correct, there are tensed facts and temporal becoming. In that case God, in virtue of His omniscience and creative activity, must know tensed facts and be the cause of things' coming to be. But in doing those things, God changes both extrinsically and intrinsically and therefore must be temporal.

The crucial assumption here is that a dynamic theory of time is true. We therefore devoted two chapters to an exploration of arguments for and against the dynamic and static theories of time. In favor of a dynamic theory of time are the facts that we experience tense and temporal becoming in a variety of ways and that there seem to be tensed facts, as disclosed by the ineliminability of tense from language. The objections typically lodged against a dynamic theory of time are really aimed at a straw man, a sort of hybrid theory according to which all events in time are equally real and "presentness" moves along the series of events. On the other hand, the principal arguments in favor of a static theory of time from Relativity Theory and the so-called mind-dependence of becoming are based upon a fundamentally flawed understanding of time which collapses time to our physical measures of time—a reductionism which theists have every reason to reject. Moreover, powerful philosophical and theological objections stand against the static theory of time. The most plausible view of the nature of time, then, is that time involves an objective distinction between past, present, and future, and that temporal becoming is a real, mind-independent feature of the world.

It therefore follows from our arguments that God is (present tense) in time. He exists now. But on the Christian doctrine of creation, the world had a beginning, though God did not. Did time exist prior to the moment of creation? Is God, existing alone without creation, timeless or temporal in such a state? I presented three arguments to show that (metric) time is finite in the past, so that God existing without the world must exist either in an amorphous time or, more plausibly, timelessly. In short, given the reality of tense and temporal becoming, the most plausible construal of divine eternity is that God is timeless without creation and temporal since creation.

This remarkable conclusion merits our reflection. Like the incarnation, the creation of the world is an act of condescension on God's part for the sake of His creatures. Alone in the self-sufficiency of His own being, enjoying the timeless fullness of the intra-Trinitarian love relationships, God had no need for the creation of finite persons. His timeless, free decision to create a temporal world with a beginning is a decision on God's part to abandon timelessness and to take on a temporal mode of existence. He did this, not out of any deficit in Himself or His mode of existence, but in order that finite temporal creatures might come to share the joy and blessedness of the inner life of God. He stooped to take on a mode of existence inessential to His being or happiness in order that we might have being and find supreme happiness in Him. His taking a human nature into intimate union with Himself in the incarnation of the *Logos*, the second person of the Trinity, was thus not what the Danish philosopher Kierkegaard regarded as "the Absurd," the union of eternity with time, for God was already temporal at the time of the incarnation and had been since the inception of creation. But in the incarnation God stooped even lower to take on, not just our mode of existence, our temporality, but our very nature.

As a result of God's creation of and entry into time, He is now with us literally moment by moment as we live and breathe, sharing our every second. He is and will be always with us. He remembers all that has transpired, knows all that is happening, and foreknows all that is to come, not only in our individual lives but throughout the entire universe. Unfettered by the finite velocity of light and clock synchronization procedures, He is, as Newton said, the Lord God of dominion throughout His universe. Well did St. Jude exclaim, "To the only God our Savior through Jesus Christ our Lord, be glory, majesty, dominion, and authority, before all time, and now and forever! Amen!"

APPENDIX

DIVINE ETERNITY AND GOD'S KNOWLEDGE OF THE FUTURE

Introduction

In the present work I have argued that divine eternity is most plausibly construed in terms of God's timelessness without creation and His temporality since the moment of creation. Now this view raises all sorts of interesting questions concerning divine omniscience. It implies that since the moment of creation God possesses literal foreknowledge, rather than timeless knowledge, of events which will happen in the future. Indeed, it will be recalled that God's knowledge of tensed facts provided one of the principal reasons for thinking God to be temporal rather than timeless.[1] As an omniscient being, God cannot be ignorant of future-tense facts. For example, prior to December 7, 1941, He knew the tensed fact *The Japanese will attack Pearl Harbor on December 7, 1941*, just as subsequent to that date He knew the tensed fact *The Japanese attacked Pearl Harbor on December 7, 1941*. In virtue of His knowing all future-tense facts, as well as all present- and past-tense facts, God has literal foreknowledge of the future.

The doctrine of divine foreknowledge has been challenged principally on two grounds: (1) It is alleged that if God foreknows future events, then those events will necessarily occur, which precludes human freedom; and (2) it is claimed that if events do occur contingently, then such events cannot be known in advance. It is often said that God, therefore, does not know which future contingent events will transpire and that He is thus a "risk-taking" God. Sometimes it is claimed that such a conception of God is faithful to biblical teaching.

I have dealt with both of these challenges to divine foreknowledge in my

---
[1] See chapter 3, pages 97-109.

book *The Only Wise God,* where I attempt to show that there is no incompatibility between God's foreknowledge and human freedom and that foreknowledge of a contingent future is possible.[2] That book was, however, somewhat ahead of the curve, and since its publication the debate over divine foreknowledge has become white-hot among Christian theologians. Therefore, I think it appropriate to re-address the question in this place.

The Biblical Doctrine of Divine Foreknowledge

The suggestion that the God described in the biblical tradition is ignorant of future contingents is on the face of it an extraordinary claim. For not only are the Scriptures replete with examples of precisely such knowledge on God's part, but they explicitly teach that God has foreknowledge of future events, even employing a specialist vocabulary to denominate such knowledge. The New Testament introduces a whole family of words associated with God's knowledge of the future, such as "foreknow" (*proginōskō*), "foreknowledge" (*prognōsis*), "foresee" (*prooraō*), "foreordain" (*proorizō*), and "foretell" (*promarturomai, prokatangellō*). Thus the claim that the biblical concept of omniscience does not comprise knowledge of the future seems frivolous.

The affirmation of God's knowledge of the future is important in two respects. First, this aspect of divine omniscience underlies the biblical scheme of history. For the biblical conception of history is not that of an unpredictably unfolding sequence of events plunging haphazardly without purpose or direction; rather God knows the future and directs the course of world history toward His foreseen ends:

> I am God, and there is none like me,
> declaring the end from the beginning
>   and from ancient times things not yet done,
> saying, "My counsel shall stand,
>   and I will accomplish all my purpose" (Isa. 46:9-10).

Biblical history is a salvation history, and Christ is the beginning, centerpiece, and culmination of that history. God's salvific plan was not an afterthought necessitated by an unforeseen circumstance. Paul speaks of "the plan of the mystery hidden for ages in God who created all things," "a plan for the fullness of time" according to "the eternal purpose which he has realized in Christ Jesus our Lord" (Eph. 3:9; 1:10; 3:11; cf. 2 Tim. 1:9-10). Similarly,

---

[2] William Lane Craig, *The Only Wise God* (Grand Rapids, Mich.: Baker, 1987; rep. ed.: Eugene, Ore.: Wipf & Stock, 2000).

Peter states that Christ "was destined before the foundation of the world but was made manifest at the end of the times for your sake" (1 Pet. 1:20). God's knowledge of the course of world history and His control over it to achieve His purposes are fundamental to the biblical conception of history and are a source of comfort and assurance to the believer in times of distress.

Second, God's knowledge of the future seems essential to the prophetic pattern that underlies the biblical scheme of history. The test of the true prophet was success in foretelling the future: "When a prophet speaks in the name of the LORD, if the word does not come to pass or come true, that is a word which the LORD has not spoken" (Deut. 18:22). The history of Israel was punctuated with prophets who foretold events in both the immediate and distant future, and it was the conviction of the New Testament writers that the coming and work of Jesus had been prophesied.

The prophetic element, however, is not limited to the fulfillment of Old Testament predictions. Jesus himself is characterized as a prophet, and he predicts the destruction of Jerusalem, signs of the end of the world, and his own return as Lord of all nations (Matt. 24; Mark 13; Luke 21). In the early church, too, there were prophets who told of events to come (Acts 11:27-28; 21:10-11; see also 13:1; 15:32; 21:9; 1 Cor. 12:28-29; 14:29, 37; Eph. 4:11). The Revelation to John is a mighty vision of the end of human history: ". . . the Lord, the God of the spirits of the prophets, has sent his angel to show his servants what must soon take place" (Rev. 22:6). The prophetic pattern thus reveals an underlying unity, not only between the two Testaments but beneath the entire course of human history.

The biblical view of history and prophecy thus seems to necessitate a God who knows not only the present and past but also the future. Indeed, so essential is God's knowledge of the future that Isaiah makes knowledge of the future the decisive test in distinguishing the true God from false gods. The prophet flings this challenge in the teeth of all pretenders to deity:

> Set forth your case, says the LORD;
>   bring your proofs, says the King of Jacob.
> Let them bring them, and tell us
>   what is to happen.
> Tell us the former things, what they are
>   that we may consider them,
> that we may know their outcome;
>   or declare to us the things to come.
> Tell us what is to come hereafter,

> that we may know that you are gods;
> do good, or do harm,
> > that we may be dismayed and terrified.
> Behold, you are nothing,
> > and your work is naught;
> > an abomination is he who chooses you (Isa. 41:21-24).

Stephen Charnock in his classic *Existence and Attributes of God* comments on this passage:

> Such a foreknowledge of things to come is here ascribed to God by God himself, as a distinction of him from all false gods. Such a knowledge that, if any could prove that they were possessors of, he would acknowledge them as gods as well as himself: "that we may know that you are gods." He puts his Deity to stand or fall upon this account, and this should be the point which should decide the controversy whether he or the heathen idols were the true God. The dispute is managed by this medium: he that knows things to come is God; I know things to come, *ergo* I am God: the idols know not things to come, therefore they are not gods. God submits the being of his Deity to this trial. If God knows things to come no more than the heathen idols, which were either devils or men, he would be, in his own account, no more a God than devils or men. . . . It cannot be understood of future things in their causes, when the effects necessarily arise from such causes, as light from the sun and heat from the fire. Many of these men know; more of them, angels and devils know; if God, therefore, had not a higher and farther knowledge than this, he would not by this be proved to be God, any more than angels and devils, who know necessary effects in their causes. The devils, indeed, did predict some things in the heathen oracles, but God is differenced from them here . . . in being able to predict things to come that they knew not, or things in their particularities, things that depended on the liberty of man's will, which the devils could lay no claim to a certain knowledge of. Were it only a conjectural knowledge that is here meant, the devils might answer they can conjecture, and so their deity were as good as God's. . . . God asserts his knowledge of things to come as a manifest evidence of his Godhead; those that deny, therefore, the argument that proves it, deny the conclusion, too; for this will necessarily follow, that if he be God because he knows future things, then he that doth not know future things is not God; and if God knows not future things but only by conjecture, then there is no God, because a certain knowledge, so as infallibly to predict things to come, is an inseparable perfection of the Deity.[3]

---

[3] Stephen Charnock, *The Existence and Attributes of God* (1682; reprint, Grand Rapids, Mich.: Baker, 1979), vol. 1, 431-432.

As Charnock notes, God's knowledge seems to encompass future contingencies. Just as God knows the thoughts humans have, so he foreknows the very thoughts they will have. The psalmist declares,

> O LORD, thou hast searched me and known me!
> Thou knowest when I sit down and when I rise up;
>   thou discernest my thoughts from afar.
> Thou searchest out my path and my lying down,
>   and art acquainted with all my ways.
> Even before a word is on my tongue,
>   lo, O LORD, thou knowest it altogether.
> Thou dost beset me behind and before,
>   and layest thy hand upon me.
> Such knowledge is too wonderful for me;
>   it is high, I cannot attain it (Ps. 139:1-6).

Here the psalmist envisages himself as surrounded by God's knowledge. God knows everything about him, even his thoughts. "From afar" (*mērāḥôq*) may be taken to indicate temporal distance—God knows the psalmist's thoughts long before he thinks them. Similarly, even before he speaks a word, God knows what he will say. Little wonder that such knowledge is beyond the reach of the psalmist's understanding! But such is the knowledge of Israel's God in contradistinction to all the false gods of her neighbors. The God of Israel was conceived to possess knowledge of the future, a property which distinguished Him from all false gods.

In light of the clear biblical affirmations of divine foreknowledge, it might seem remarkable that some otherwise conservative theologians would deny that the Bible teaches that God knows future events. They argue that God can only make intelligent conjectures about what free creatures are going to do. As a result God is ignorant of vast stretches of human history, since even a single free choice could divert history from its present course, and subsequent events would, as time goes on, depart increasingly from history's present trajectory. At best God can be said to have a good idea of what will happen only in the very near future.

Such a view seems so unbiblical that we might be surprised to hear that some persons think that it represents faithfully the doctrine of the Scriptures. Those who hold to this view, however, typically point to passages in the Scriptures which imply that God is ignorant of some fact (Jer. 26:3; 36:3). The problem with trying to base a doctrine of divine omniscience on such passages, however, is that it underestimates the degree to which the narratives of

God's acts are anthropomorphic in character; that is to say, God is described in human terms which are not intended to be taken literally. The Bible is not a treatise in theology, much less philosophy of religion. It is a collection of stories about God's dealings with His people. The storyteller's art is not to reflect philosophically upon his narrative but to tell a vivid tale. Thus, the Scriptures are filled with anthropomorphisms, many so subtle that they escape our notice. There are not only the obvious anthropomorphisms, such as references to God's eyes, hands, and nostrils, but almost unconscious anthropomorphisms, such as references to God's seeing the distress of His people, hearing their prayers, crushing His enemies, turning away from apostate Israel, and so forth. These are all of them *metaphors,* since God does not possess literal bodily parts. In the same way, given the explicit teaching of Scripture that God does foreknow the future, the passages which portray God as ignorant or inquiring are probably just anthropomorphisms characteristic of the genre of narrative.

Those who deny divine foreknowledge also appeal to passages in which God predicts that something will happen, but then repents, so that the predicted event does not come to pass (Amos 7:1-6; Jonah 3; Isa. 38:1-5). Obviously, since what God predicted did not in the end happen, the predictions were not foreknowledge of the future. The problem here is how to explain that, while the authors of these passages were aware that God knew the future and could not lie (Num. 23:19; 1 Sam. 15:29), yet they represent Him as relenting on impending judgments which He had commanded His prophets to proclaim.

The most plausible interpretation of such passages is that these prophecies were not simple glimpses of the future but pictures of what was going to happen *unless . . .* [4] The prophecies contained the implicit condition "all things remaining the same." Certain prophecies thus are forecasts or forewarnings of what is going to happen if all things remain as they are. Such events are sometimes referred to as conditional future contingents, and God's knowledge of such events is even more remarkable than simple foreknowledge, since it involves knowledge of what would happen were other circumstances to exist than those that will. Not all of the prophecies in the Old and New Testaments are forewarnings, however. Prophecies of events which are brought about not by God but by human beings and which could not have been inferred from present causes cannot be interpreted as forewarnings but must be considered to express simple foreknowledge on God's part.

---

[4] Witherington calls these conditional prophecies (Ben Witherington III, *Jesus the Seer* (Peabody, Mass.: Hendrickson, 1999), 3; cf. 134.

How do the detractors of foreknowledge explain scriptural passages which illustrate God's knowledge of the future? Typically, they attempt to dismiss each example of divine foreknowledge as being one of the following: (1) a declaration by God of what He Himself intends to bring about, (2) an inference of what is going to happen based on present causes, (3) a conditional prediction of what will happen *if* something else happens.

Such an account seems inadequate, however. As far as (3) is concerned, conditional predictions, if they do not reduce to (1) or (2), must be expressions of what theologians call divine middle knowledge, which is even more remarkable than divine foreknowledge and, indeed, may provide the basis for divine foreknowledge.[5] Hence, to try to explain away divine foreknowledge by means of (3) is counterproductive.

As for (2), while it might be claimed, say, that Jesus predicted Judas's betrayal or Peter's denial solely on the basis of their character and the surrounding circumstances, there can be no question that the Gospel writers themselves did not so understand such predictions. To try to explain biblical prophecies as mere inferences from present states of affairs denudes them of any theological significance. The writers of Scripture clearly saw prophecy not as God's reasoned conjecture of what will happen but as a manifestation of His infinite knowledge, encompassing even things yet to come.

As for (1), it is true that many prophecies in Scripture are clearly based on God's irrevocable intention to bring about certain future events on His own. In such cases, prophecy serves to manifest not so much God's omniscience as His omnipotence, His ability to bring about whatever He intends. But the problem with (1) is that it simply cannot be stretched to cover all the cases. Divine foreknowledge of free human actions cannot be accounted for by (1), since it negates human freedom. Explanation (1) is useful only in accounting for God's knowledge of events which He Himself will bring about. But the Scripture provides many examples of divine foreknowledge of events which God does not directly cause, events which are the result of free human choices.

Finally, none of the three explanations comes to grips with the Scriptures' doctrinal teaching concerning God's foreknowledge. These explanations try to account only for examples of prophecy in the Bible and say nothing about the passages which explicitly teach that God foreknows the future. Thus we

---

[5] Middle knowledge involves knowledge of subjunctive conditionals such as *If Goldwater had been elected in 1964, he would have won the Vietnam War.* See my article "Middle Knowledge," in *Four Views on Divine Knowledge*, ed. James Beilby and Paul Eddy (Downer's Grove, Ill.: InterVarsity, forthcoming).

have strong biblical warrant for the doctrine that God's omniscience encompasses knowledge of future contingents.

Philosophical Grounds for Affirming Divine Foreknowledge

Not only are there biblical grounds for affirming God's foreknowledge of future contingents, but there are good philosophical reasons for thinking that God foreknows the future. As St. Anselm saw, the concept of God is the concept of a perfect being, what Anselm termed the greatest conceivable being. (Just ask whether any being which is less than perfect would be worthy of worship.) Now the greatest conceivable being, a perfect being, must be all-knowing or omniscient. For ignorance is an imperfection; all things being equal, it is greater or better to be knowledgeable than ignorant. Therefore, if there are truths about future contingents, God, as an omniscient being, must know these truths. Since there are such truths about the future, that is to say, since statements about future contingents are either true or false, and they are not all false, God must therefore know all truths about the future, which is to say that He knows future-tense facts; He knows what will happen.

One might try to escape the force of this reasoning by contending that future-tense statements are neither true nor false, so that there are no facts about the future.[6] Such a view cannot, however, be plausibly maintained. Here several points deserve mention:

First, there is no good reason to deny that future-tense statements are either true or false. Why should we accept the view that future-tense statements about free acts, statements which we make all the time in ordinary conversation, are in fact neither true nor false? What proof is there that such statements are neither true nor false?

About the only answer of any substance ever given to this question goes something like this: Future events, unlike present events, do not exist. Now, a statement is true if and only if it corresponds to what exists, and false if and only if it does not correspond to what exists. Since the future does not exist, there is nothing for future-tense statements to correspond

---

[6] It is very important here that we realize that by a "statement" I do not mean a sentence token (recall the distinction between sentence tokens and types made in chapter 4, note 4). Otherwise we should have to say that during the Jurassic Age, when no human beings were about, there were no future-tense statements (and so no facts about the future), but that now there are! Rather by "statement" I mean something more like a sentence type, which may or may not ever be uttered or written. So, for example, even though the sentence token "No sentences exist" can never be true, clearly the statement that *No sentences exist* can be and often has been true. The question before us does not concern whether there happen to be true future-tense sentences either uttered or written, but whether there are truths about the future, and so I use the term "statement" to ask this question.

with or to fail to correspond with. Hence, future-tense statements cannot be true or false.

Since I accept the view of time which this proposed answer presupposes (namely, the dynamic theory of time), the issue is whether, given such a view, the idea of truth as correspondence requires us to deny that future-tense statements are either true or false. Those who think that it does seem to misunderstand the concept of truth as correspondence, which holds merely that a statement is true if and only if what it states to be the case really is the case. For example, the statement "It is snowing" is true if and only if it is snowing. Although this might seem too obvious to be worth stating, it is sometimes misunderstood. Truth as correspondence does *not* mean that the things or events which a true statement is about must exist. Indeed, it is only in the case of true present-tense statements that the things or events referred to must exist. For a past-tense statement to be true it is not required that what it describes exist, but only that it *have* existed. For a future-tense statement to be true it is not required that what it describes exist, but that it *will* exist. In order for a future-tense statement to be true, all that is required is that when the moment described arrives, the present-tense version of the statement will be true at that moment. The idea that the concept of truth as correspondence requires that the things or events described by the statement must exist at the time the statement is true is a complete misunderstanding.

To say that a future-tense statement is now true is not, of course, to say that we may now know whether it is true or to say that things are now so determined that it is true. It is only to say that when the time arrives, things will turn out as the statement predicts. A future-tense statement is true if matters turn out as the statement predicts, and false if matters fail to turn out as the statement predicts—this is all that the notion of truth as correspondence requires. Hence, there is no good reason to deny that future-tense statements are either true or false.

Second, there are good reasons to maintain that future-tense statements are either true or false.

(i) The same facts that guarantee the truth or falsity of present- and past-tense statements also guarantee the truth or falsity of future-tense statements. Nicholas Rescher explains,

> Difficulties about divine foreknowledge quite apart, it is difficult to justify granting to
> 1. "It will rain tomorrow" (asserted on April 12)
> a truth status different from that of

2. "It did rain yesterday" (asserted on April 14)

because both make (from temporally distinct perspectives) *precisely the same claim about the facts,* viz., rain on April 13.[7]

Think about it for a moment. If "It is raining today" is now true, how could "It will rain tomorrow" not have been true yesterday? The same facts guarantee that a future-tense statement asserted earlier, a present-tense statement asserted simultaneously, and a past-tense statement asserted later are all true.

(ii) If future-tense statements are not true, then neither are past-tense statements true. If future-tense statements cannot be true because the realities they describe do not yet exist, then by the same token past-tense statements cannot be true because the realities they describe no longer exist. But to maintain that past-tense statements cannot be true would be ridiculous. Since the two cases are parallel, one must either deny the truth or falsity of both past- and future-tense statements or affirm the truth or falsity of both.

(iii) Tenseless statements are always true or false. Recall that it is possible to eliminate the tense of the verb in a statement and specify the time at which the statement is supposed to be true.[8] For example, the statement "The Allies invaded Normandy" can be made tenseless by specifying the time: "On June 6, 1944, the Allies *invade* Normandy," the italics indicating that the verb is tenseless. If the tensed version is true, then so is the tenseless version.[9] Thus, correlated with any true past- or present-tense statement is a true tenseless version of that statement. Furthermore, a tenseless statement, if it is true at all, is *always* true. This is precisely because the statement is tenseless. If "On June 6, 1944, the Allies *invade* Normandy" is *ever* true, then it is *always* true. Therefore, this statement is true prior to June 6, 1944. But in that case, it is true prior to June 6, 1944, that the Allies on that date will invade Normandy, which is the same as saying that the future-tense version of the statement is true. Moreover, since God is omniscient, He must always know the truth of the tenseless statement, which entails that He foreknows the future.

Third, the denial of the truth or falsity of future-tense statements has absurd consequences. For example, if future-tense statements are neither true nor false, the statement made in 1998 "George W. Bush either will or will not win the presidential election in 2000" would not be true. For this statement

---

[7] Nicholas Rescher, *Many-Valued Logic* (New York: McGraw-Hill, 1969), 2-3.
[8] See chapter 3, page 99.
[9] For a good discussion see Thomas Bradley Talbott, "Fatalism and the Timelessness of Truth" (Ph.D. diss., University of California at Santa Barbara, 1974), 153-154.

is a compound made up of two simple future-tense sentences—"George W. Bush will win the presidential election in 2000" and "George W. Bush will not win the presidential election in 2000." And if neither of these individual statements is true or false, the compound statement combining them is also neither true nor false. But how can this be? Either Bush will win or he will not—there is no other alternative. But the view that future-tense statements are neither true nor false would require us to say that this compound statement is neither true nor false, which seems absurd.

Equally absurd, we could not say that a statement such as "Bush both will and will not win the presidential election in 2000" is false. For this is a compound statement consisting of two simple future-tense statements, neither of which is supposed to be true or false. Therefore, the compound statement cannot be true or false either. But surely this statement is false, for it is a self-contradiction: Bush cannot both win and not win the election!

We must conclude that with no good reason in favor of it, persuasive reasons against it, and absurd consequences following from it, the view that future-tense statements about free acts are neither true nor false is untenable. The view that God's omniscience does not encompass foreknowledge is thereby seen to be untenable, since as an omniscient being He must know all true statements, including all true future-tense statements.

Detractors of divine foreknowledge often try to escape this conclusion by redefining the concept of omniscience in such a way that *being omniscient* does not entail *knowing all truths*. Thus, they must reject the usual definition of omniscience:

O. For any agent x, x is omniscient = $_{def.}$ For every statement s, if s is true, then x knows that s and does not believe that not-s.

What (O) requires is that a person is omniscient if and only if he knows all truths and believes no falsehoods. This is the standard definition of omniscience. It entails that if there are future-tense truths, then an omniscient being must know them.

So as not to deny God's omniscience, opponents of divine foreknowledge have suggested revisionary definitions of omniscience in order to be able to affirm that God is omniscient even as they deny His knowledge of future contingents.[10] William Hasker's revisionist definition is typical:

---

[10] For the following definition see William Hasker, "A Philosophical Perspective," in Clark Pinnock, Richard Rice, John Sanders, William Hasker, and David Basinger, *The Openness of God: A Biblical Challenge to the Traditional Understanding of God* (Downer's Grove, Ill: InterVarsity, 1994), 136.

O'. God is omniscient = $_{def.}$ God knows all statements which are such that God's knowing them is logically possible.

Revisionists then go on to claim that it is logically impossible to know statements about future contingents, and so God may count as omniscient despite His ignorance of an infinite number of true statements.

As it stands, however, (O') is drastically flawed, for it does not exclude that God believes false statements as well as true ones. Worse, (O') actually requires God to know false statements, which is incoherent as well as theologically unacceptable. For (O') requires that if it is logically possible for God to know some statement *s*, then God knows *s*. But if *s* is a contingently false statement, say, *There are eight planets in the sun's solar system*, then there are logically possible worlds in which *s* is true and so known by God. Therefore, since it is logically possible for God to know *s*, He must according to (O') actually know *s*, which is absurd.

What the revisionist really wants to say is something like

O''. God is omniscient = $_{def.}$ God knows only and all true statements which are such that it is logically possible for God to know them.

Unlike (O'), (O'') limits God's knowledge to a certain subset of all true statements.

The fundamental problem with all such revisionary definitions of omniscience as (O'') is that, as we have seen[11], any adequate definition of a concept must accord with our intuitive understanding of the concept. We are not at liberty to "cook" the definition in some desired way without thereby making the definition unacceptably contrived. (O'') is guilty of being "cooked" in this way. For, intuitively, omniscience involves knowing all truths, yet according to (O'') God could conceivably be ignorant of infinite realms of truths and yet still count as omniscient. The only reason why someone would prefer (O'') to (O) is due to an ulterior motivation to salvage the attribute of omniscience for a cognitively limited deity rather than to deny outright that God is omniscient. (O'') is therefore unacceptably contrived.

A second problem with (O'') is that it construes omniscience in modal terms, speaking, not of knowing all truth, but of knowing all truth which is knowable. But omniscience, unlike omnipotence, is not a modal notion. Roughly speaking, omnipotence is the capability of actualizing any logically

---
[11] See chapter 3, page 104.

possible state of affairs. But omniscience is not merely the *capability* of knowing only and all truths; it *is* knowing only and all truths. Nor does omniscience mean knowing only and all knowable truths, but knowing only and all truths, period. It is a categorical, not a modal, notion.

Third, the superiority of (O'') over (O) depends on there being a difference between a truth and a truth which it is logically possible to know. If there is no difference, then (O'') collapses back to (O), and the revisionist has gained nothing. But it is far from evident that there is any difference. For what is a sufficient condition for a statement to be logically knowable? So far as I can see, the only condition is that the statement be true. What more is needed? If the revisionist thinks that something more is needed, then we may ask him for an example of a statement that could be true but logically impossible to know. A statement such as "Nothing exists" or "All agents have ceased to exist" comes to mind; but on traditional theism these statements are not possibly true, since God is an agent whose non-existence is impossible. Unless the revisionist can give us some reason to think that a statement can be true yet unknowable, we have no reason to adopt (O''). It seems that the only intrinsic property which a statement must possess in order to be logically knowable is truth.

The revisionist will claim at this point that future contingent statements are logically impossible for God to know, since if He knows them, then they are not contingent.[12] We shall examine the revisionist's argument for this latter claim below; but here we may note that even if we concede that his argument is sound, it still does not follow that future contingent statements are logically impossible for God to know. The revisionist reasons that for any future-tense statement s it is impossible that God know s and that s be contingently true; therefore, if s is contingently true, it is not possible that God knows s. But such reasoning is logically fallacious. From

1. Not-possibly (God knows s, and s is contingently true)

and

2. s is contingently true

it does not follow logically that

3. Not-possibly (God knows s)

---

[12] Hasker, "Philosophical Perspective," 147-148.

but merely

> 3'. Not (God knows *s*).

In other words, what follows from (1) and (2) is merely that God does not know *s*, not that it is impossible that God knows *s*. Thus, even *granted* the revisionist's premise that it is impossible that God know *s* and *s* be contingently true, it does not follow from the contingent truth of *s* that *s* is such that it is logically impossible for God to know *s*. Therefore, even on the defective definition (O'') proposed by the revisionist, the revisionist's God turns out not to be omniscient, since *s* is a true statement which, so far as we can see, is logically possible for God to know, and yet God does not know *s*. Thus, the revisionist must deny divine omniscience and therefore reject God's perfection—a very serious theological consequence, indeed.[13]

Philosophical Objections to Divine Foreknowledge

As I mentioned, opponents of divine foreknowledge usually raise two objections to that doctrine: (1) Divine foreknowledge implies fatalism, and (2) there is no basis on which God can know future contingents. Let us explore each of these issues in turn.

The Compatibility of Divine Foreknowledge and Future Contingents

The first objection raises the issue of fatalism, the doctrine that everything we do we do necessarily and that therefore human freedom is an illusion. It is alleged that if God foreknows the future, then fatalism is true. Since fatalism is not true, it follows that God must not foreknow the future.

What is the argument that allegedly demonstrates the connection between divine foreknowledge and fatalism? Letting "*x*" stand for any event, the basic form of the argument is as follows:

> 1. Necessarily, if God foreknows *x*, then *x* will happen.

---

[13] Notice, too, that the revisionist's position is ultimately logically incoherent. For by his own lights it is logically possible to know any true, present-tense statement. But if future-tense statements are true or false, then there will be present-tense statements such as "Future-tense statement *s* is presently true" which must be known to God. It cannot reasonably be denied that God must know such present-tense statements, for God knows what properties presently inhere in existing things. But then He must know that "Truth presently inheres in future-tense statement *s*." Hence, the detractor of divine foreknowledge cannot coherently affirm that there are true future-tense statements and yet deny that God knows such statements—he must deny the truth or falsity of future-tense statements, a radical position.

2. God foreknows *x*.

3. Therefore, *x* will necessarily happen.

Since *x* happens necessarily, it is not a contingent event. In virtue of God's foreknowledge, everything is fated to occur.

The problem with the above form of the argument is that it is just logically fallacious. What is validly implied by premises (1) and (2) is not (3) but

3'. Therefore, *x* will happen.

The fatalist gets things all mixed up here. It is correct that in a valid, deductive argument the premises necessarily imply the conclusion. The conclusion follows necessarily from the premises; that is to say, it is impossible for the premises to be true and the conclusion to be false. But the conclusion itself need not be necessary. The fatalist illicitly transfers the necessity of the *inference* to the conclusion *itself*. What necessarily follows from (1) and (2) is just (3'). But the fatalist in his confusion thinks that the conclusion is itself necessarily true and so winds up with (3). In so doing he simply commits a common logical fallacy.

The correct conclusion (3') is in no way incompatible with human freedom. From God's knowledge that I shall do *x*, it does not follow that I must do *x* but only that I shall do *x*. That is in no way incompatible with my doing *x* freely.

Undoubtedly a major source of the fatalist's confusion is his conflating *certainty* with *necessity*. One frequently finds in the writings of contemporary theological fatalists statements which slide from affirming that something is *certainly* true to affirming that it is *necessarily* true. This is sheer confusion. Certainty is a property of persons and has nothing to do with truth, as is evident from the fact that we can be absolutely certain about something which turns out to be false. (Dogmatic people often have this problem.) By contrast, necessity is a property of statements, indicating that a statement cannot possibly be false. We can be wholly uncertain about statements which are, unbeknownst to us, necessarily true (imagine some complex mathematical equation or theorem). Thus, when we say that some statement is "certainly true," this is but a manner of speaking indicating that we are certain that the statement is true. People are certain; statements are necessary.

By confusing certainty and necessity, the fatalist makes his logically fallacious argument deceptively appealing. For it is correct that from premises

(1) and (2) we can be absolutely certain that x will come to pass. But it is muddle-headed to think that, because x will certainly happen, x will necessarily happen. We can be certain, given God's foreknowledge, that x will not fail to happen, even though it is entirely possible that x fail to happen. X could fail to occur, but God knows that it will not. Therefore, we can be sure that it will happen—and happen contingently.

Contemporary theological fatalists recognize the fallaciousness of the above form of the argument and therefore try to remedy the defect by making premise (2) also necessarily true:

1. Necessarily, if God foreknows x, then x will happen.

2'. Necessarily, God foreknows x.

3. Therefore, x will necessarily happen.

So formulated, the argument is no longer logically fallacious, and so the question becomes whether the premises are true.

Premise (1) is clearly true. It is perhaps worth noting that this is the case, not because of God's essential omniscience or inerrancy, but simply in virtue of the definition of "knowledge." Since knowledge entails true belief, anybody's knowing that x will happen implies necessarily that x will happen. Thus, we could replace (1) and (2') with

1.* Necessarily, if Smith truly believes that x will happen, then x will happen.

2.* Necessarily, Smith truly believes that x will happen.

And (3) will follow as before. Therefore, if any person ever holds true beliefs about the future (and surely we do, as we smugly remind others when we say, "I told you so!"), then, given the truth of premise (2*), fatalism follows from merely human beliefs, a curious conclusion!

Indeed, as ancient Greek fatalists realized, the presence of any agent at all is really superfluous to the argument. All one needs is a true, future-tense statement to get the argument going. Thus, we could replace (1) and (2') with

1.** Necessarily, if it is true that x will happen, then x will happen.

2.** Necessarily, it is true that x will happen.

And we shall get (3) as our conclusion. Thus, philosopher Susan Haack quite rightly calls the argument for theological fatalism "a needlessly (and confusingly) elaborated version" of Greek fatalism; the addition of an omniscient God to the argument constitutes a "gratuitous detour" around the real issue, which is the truth or falsity of future-tense statements.[14]

In order to avoid the above generalization of their argument to all persons and to mere statements about the future, theological fatalists will deny that the second premise is true with respect to humans or mere statements, as it is for God. They will say that Smith's holding a true belief or some future-tense statement's being true are not necessary in the way that God's holding a belief is necessary.

That raises the question as to whether premise (2') is true. Now at face value, premise (2') is obviously false. Christian theology has always maintained that God's creation of the world is a free act, that God could have created a different world, in which x does not occur, or even no world at all. To say that God necessarily foreknows any event x implies that this is the only world God could have created and thus denies divine freedom.

But theological fatalists have a different sort of necessity in mind when they say that God's foreknowledge is necessary. What they are talking about is *temporal necessity*, or the necessity of the past. Often this is expressed by saying that the past is unpreventable or unchangeable. If some event is in the past, then it is now too late to do anything to affect it. It is in that sense necessary. Since God's foreknowledge of future events is now part of the past, it is now fixed and unalterable. Therefore, it is said, premise (2') is true.

But if premise (2') is true in that sense, then why are not (2*) and (2**) true as well? The theological fatalist will respond that Smith's belief's being true or a future-tense statement's being true are not facts or events of the past, as is God's holding a belief.

But such an understanding of what constitutes a fact or event seems quite counterintuitive. If Smith believed in 1997 that "Bill Clinton will be impeached," was it not a fact that his belief was true? If Smith held that same belief today, would it not be a fact that his belief is no longer true (since Clinton no longer holds office)? If Smith's belief thus changed from being true to being false, then surely it was a fact that it was then true and is a fact that it is now false. The same obviously goes for the mere statement "Bill Clinton

---

[14] Susan Haack, "On a Theological Argument for Fatalism," *Philosophical Quarterly* 24 (1974): 158.

will be impeached." This statement once had the property of being true and now has the property of being false. In any reasonable sense of "fact," these are past and present facts.

Indeed, a statement's having a truth value is plausibly an event as well. This is most obvious with respect to statements such as "Flight 4750 to Paris will depart in five minutes." That statement is false up until five minutes prior to departure, becomes true at five minutes till, and then becomes false again immediately thereafter. Other statements' being true may be more long-lasting events, such as "Flight 4750 to Paris will depart within the next hour." Such statements' being true are clearly events on any reasonable construal of what constitutes an event.

No theological fatalist that I have read has even begun to address the question of the nature of facts or events which would make it plausible that Smith's truly believing a future-tense statement and a future-tense statement's being true do not count as past facts or events. But then we see that theological fatalism is not inherently theological at all; if the theological fatalist's reasoning is correct, it can be generalized to show that every time we hold a true belief about the future or even every time a statement about the future is true, then the future is fated to occur—surely an incredible inference!

Moreover, we have the best of reasons for thinking that premise (2') is defective in some way, namely, fatalism posits a constraint on human freedom which is unintelligible. For the fatalist admits that the events foreknown by God may be causally indeterminate; indeed, they could theoretically be completely uncaused, spontaneous events. Nevertheless, such events are said to be somehow constrained. But by what? By fate? What is that but a mere name? If my action is causally free, how can it be constrained by the mere fact of God's knowing about it?

Sometimes fatalists say that God's foreknowledge places a sort of logical constraint on my action. Even though I am causally free to refrain from my action, there is some sort of logical constraint upon me, rendering it impossible for me to refrain. But insofar as we can make sense of logical constraints, they are not analogous to the sort of necessitation imagined by the theological fatalist. For example, given the fact that I have already played basketball at least once in my life, it is now impossible for me to play basketball for the first time. I am thus not free to go out and play basketball for the first time. But this sort of constraint is not at all analogous to theological fatalism. For in the case we are envisioning, it is within my power to play basketball or not. Whether I have played before or not, I can freely execute the actions of playing basketball. It is just that if I have played before, my actions will not *count* as playing for the first time. By contrast the fatalist imagines that if God knows that I shall not play basketball, then even though I am causally free, my actions are mysteriously constrained so that I am literally unable to walk out onto the court, dribble, and shoot. But such non-causal determinism is utterly opaque and unintelligible.

The argument for fatalism therefore must be unsound. Since premise (1) is clearly true, the trouble must lie with premise (2'). And premise (2') is notoriously problematic, for the notion of temporal necessity appealed to by the fatalist is so obscure a concept that (2') becomes a veritable mare's nest of philosophical difficulties. For example, since the necessity of premise (1) is logical necessity and the necessity of premise (2') is temporal necessity, why think that such mixing of different kinds of modality is valid? If the fatalist answers that logical necessity entails temporal necessity, so that premise (1) can be construed merely in terms of temporal necessity, then how do we know that such necessity is passed on from the premises to the conclusion, in the way that logical necessity is? Indeed, since x is supposed to be a future event, how *could* it be temporally necessary? Since x is neither present nor past but has yet to occur, it could not possibly be characterized by the temporal necessity that is supposed to inhere in events once they have occurred. Thus, we have every reason to think that temporal necessity is not transitive.

And even if this peculiar sort of necessity were transitive and so x were temporally necessary, how do we know that this sort of necessity is incompatible with an action's being free? It is plausible that so long as a person's choice is causally undetermined, it is a free choice even if he is unable to choose the opposite of that choice.[15] Imagine a man with electrodes secretly implanted in his brain who is presented with the choice of doing either A or B. The electrodes are inactive so long as the man chooses A; but if he were going to choose B, then the electrodes would switch on and force him to choose A. If the electrodes fire, causing him to choose A, his choice of A is clearly not a free choice. But suppose that the man really wants to do A and chooses it of his own volition. In that case his choosing A is entirely free, even though the man is literally unable to choose B, since the electrodes do not function at all and so have no effect on his choice of A. What makes his choice free is the absence of any causally determining factors of his choosing A. This conception of libertarian freedom has the advantage of explaining how it is

---

[15] See Harry Frankfurt, "Alternative Possibilities and Moral Responsibility," *Journal of Philosophy* 66 (1969): 829-839; Thomas V. Morris, *The Logic of God Incarnate* (Ithaca, N.Y.: Cornell University Press, 1986), 151-152. For an application to theological fatalism see David P. Hunt, "On Augustine's Way Out," *Faith and Philosophy* 16 (1999): 3-26.

that God's choosing to do good is free, even though it is impossible for God to choose sin, namely, His choosing is undetermined by causal constraints. Thus, libertarian freedom of the will does not require the ability to choose other than as one chooses. So even if x were temporally necessary, such that not-x could not occur, it is far from obvious that x would not be freely performed or chosen.

All of the above problems arise even if we concede (2') to be true. But why think that this premise is true? What is temporal necessity anyway, and why think that God's past beliefs are now temporally necessary? Theological fatalists have never provided an adequate account of this peculiar modality. I have yet to see an explanation of temporal necessity, according to which God's past beliefs are temporally necessary, which does not reduce to either the *unalterability* or the *causal closedness* of the past.

But interpreting the necessity of the past as its unalterability (or unchangeability or unpreventability) is clearly inadequate, since the future, by definition, is just as unalterable as the past. By definition the future is what will occur, and the past is what has occurred. To *change* the future would be to bring it about that an event which will occur will not occur, which is self-contradictory. It is purely a matter of definition that the past and future cannot be changed, and no fatalistic conclusion follows from this truth. We need not be able to *change* the future in order to *determine* the future. If our actions are freely performed, then it lies within our power to determine causally what the course of future events will be, even if we do not have the power to change the future.

The fatalist will insist that the past is necessary in the sense that we do not have a similar ability to causally determine the past. The non-fatalist may happily concede the point: Backward causation is impossible. But the causal closedness of the past does not imply fatalism, for freedom to refrain from doing as God knows one will do does not involve backward causation.

One may happily admit that there is nothing I can now do to cause or bring about the past. Thus I cannot cause God to have had in the past a certain belief about my future actions. But it may well lie within my power to freely perform some action A, and if A were to occur, then the past would have been different than it in fact is. Suppose, for example, that God has always believed that in the year 2001 I would accept an invitation to speak at the University of Regensburg. Let us suppose that up until the time arrives I have the ability to accept or refuse the invitation. If I were to refuse the invitation, then God would have held a different belief than the one He in fact held. For if I were to refuse the invitation, then different future-tense statements would have been true, and God, being omniscient, would have known this. Thus, He would have had different foreknowledge than that which He in fact has. Neither the relation between my action and a corresponding future-tense statement about it, nor the relation between a true future-tense statement and God's believing it, is a causal relation. Thus, the causal closedness of the past is irrelevant. If temporal necessity is merely the causal closedness of the past, then it is insufficient to support fatalism.

No fatalist, as I say, has to my knowledge explicated a conception of temporal necessity which does not amount to either the unalterability or the causal closedness of the past. Typically, they just appeal gratuitously to some sort of "Fixed Past Principle" to the effect that it is not within my power to act in such a way, that if I were to do so, then the past would have been different—which begs the question. On analyses of temporal necessity which are not reducible to either the unalterability or the causal closedness of the past, God's past beliefs always turn out *not* to be temporally necessary.[16] It is interesting that, as I have tried to show elsewhere[17], precisely parallel conclusions follow with respect to the temporal necessity of past events in cases of time travel, backward causation, precognition, and the Special Theory of Relativity, which provide intriguing analogues to the theological scenario of God's holding beliefs about future contingents.

Thus, the argument for theological fatalism is unsound. It provides no cogent basis on which to deny the biblical doctrine of divine foreknowledge.

The Basis of Divine Foreknowledge of Future Contingents

What, then, about that second question raised by divine foreknowledge, namely, the basis of God's knowledge of future contingents? Detractors of divine foreknowledge sometimes claim that because future events do not exist, they cannot be known by God. The reasoning seems to go as follows:

1. Only events which actually exist can be known by God.

2. Future events do not exist.

3. Therefore, future events cannot be known by God.

Now premise (2) is not uncontroversial. A good many physicists and

---

[16] See, for example, Alfred J. Freddoso, "Accidental Necessity and Logical Determinism," *Journal of Philosophy* 80 (1983): 257-278.
[17] Craig, *The Only Wise God*.

philosophers of time and space argue that future events do exist. They claim that the difference between past, present, and future is merely a subjective matter of human consciousness. For the people in the year 2015 the events of that year are just as real as the events of our present are for us, and for those people, it is we who have passed away and are unreal. On such a view God transcends the four-dimensional space-time continuum, and thus all events are eternally present to Him. It is easy on such a view to understand how God could therefore know events which to us are future.

Nevertheless, I have argued that such a four-dimensional view of reality faces insuperable philosophical and theological objections.[18] Therefore, I am inclined to agree with premise (2) of the above argument. So the question becomes whether there is good reason to think that premise (1) is true.

In assessing the question of how God knows which events will transpire, it is helpful to distinguish two models of divine cognition: the *perceptualist* model and the *conceptualist* model. The perceptualist model construes divine knowledge on the analogy of sense perception. God looks and sees what is there. Such a model is implicitly assumed when people speak of God's "foreseeing" the future or having "foresight" of future events. The perceptualist model of divine cognition does run into real problems when it comes to God's knowledge of the future, for, since future events do not exist, there is nothing there to perceive.[19]

By contrast, on a conceptualist model of divine knowledge, God does not acquire His knowledge of the world by anything like perception. His knowledge of the future is not based on His "looking" ahead and "seeing" what lies in the future (a terribly anthropomorphic notion in any case). Rather God's knowledge is self-contained; it is more like a mind's knowledge of innate ideas. As an omniscient being, God has essentially the property of knowing all truths; there are truths about future events; ergo, God knows all truths concerning future events.

So long as we are not seduced into thinking of divine foreknowledge on the model of perception, it is no longer evident why knowledge of future events should be impossible. A conceptualist model furnishes a perspicuous basis for God's knowledge of future contingents.

Thus, neither the problem of theological fatalism nor the question of the basis of divine foreknowledge provides adequate grounds for denying the tes-

---

[18] Recall chapters 4 and 5.

[19] Notice, however, that if we think of statements or facts as within God's purview, then even on a perceptualist model, God can know the future, for He perceives which future-tense statements have the property of truth inhering in them or which future-tense facts presently exist. Thus, by means of His perception of presently existing realities, He knows the truth about the future. Cf. Note 13, above.

timony of both Scripture and reason to the truth of the doctrine of divine omniscience and, in particular, God's knowledge of future contingents.

Conclusion

Despite the impression conveyed by certain theologians, then, divine knowledge of what will take place in the future is not a package deal with divine timelessness. It is perfectly coherent to maintain that God is, at least since the moment of creation, temporal and also that God's omniscience extends to future contingents. Indeed, it is precisely in virtue of His omniscience that God must possess foreknowledge of such events. For if there are future-tense truths, these must be known to God. But that implies both God's temporality and His knowledge of things to come. Together these features of God entail divine foreknowledge. Such foreknowledge is wholly compatible with contingency and, in particular, human freedom and is best understood in terms of a conceptualist model of divine cognition, according to which God simply possesses essentially knowledge of all truth, including truths about future contingents.

General Index

aberration of starlight, 38
absolute becoming, *see* temporal becoming
absolute motion, 34-35, 38, 41, 48-49, 57, 172, 173, 175, 176
absolute space, 33-35, 41-43, 45, 48-49, 56-58, 63, 93
absolute time, 25, 33-34, 38-39, 43-54, 63-66, 74, 77, 151, 154, 162, 174, 231
actual infinite, 221-233
aether, 37-43, 47, 51, 55-57, 93, 176
aether drag, 38
aether wind, 57
Alston, William P., 71, 113
Amundsen, D. W., 177
anisotropy of time, 160-161, 184
Anthropic Principle, 177
Aristotle, 157
Arndt, W. F., 19
arrow of time, *see* time: direction of
Arzeliès, Henri, 95, 180
A-series, 193
A-Theory, *see* dynamic theory of time
asymmetry of time, *see* time: asymmetry of
Augustine, 14, 157, 261

backward causation, 72, 236, 262-263
Balashov, Yuri, 54, 151, 169, 215
Barbour, Ian, 212
Barr, James, 17, 20
Barrow, John D., 50, 218, 219
basic beliefs, 131-136, 139
Bauer, Walter, 19
beginning of the universe, 212, 217
beginning of time, *see* time: beginning of
behavior linked to tensed beliefs, 118
Bell, J. S., 55, 56
Bell's Theorem, 54-57, 176-177
Bergson, Henri, 159
Besso, Michael, 69
Big Bang, 21, 56, 62, 64-66, 110-111, 184, 211-212, 217-219, 221, 233, 236, 237
Black, Max, 26, 27, 95, 180, 184, 185, 187, 215
Blake, R. M., 165
Boethius, 29, 30, 67
Bohr, Niels, 54-55
Bondi, Hermann, 32, 59, 64-65
Borde, Arvind, 218
Broad, C. D., 143, 148, 156-157, 164, 165, 190
Brown, J. R., 56
B-series, 193

B-Theory, *see* static theory of time
Builder, Geoffrey, 215
Buller, David J., 164
Butterfield, Jeremy, 122

Campolo, Anthony, 24
Cantor, G. N., 225
Capek, Milic, 188, 198
Carrier, Martin, 176-177
Carter, William, 216
Castañeda, Hector-Neri, 113
change
  extrinsic, 31, 87-88, 97, 111
  intrinsic, 31, 87, 97, 110, 148-149, 200-210, 215
changelessness, 32, 94
Charnock, Stephen, 246-247
Chisholm, Roderick M., 129
chronons, 158-159
Clarke, Samuel, 230
classical concept of time, 44, 66
Cleugh, Mary F., 185, 215
clock retardation, *see* time dilation
clock synchronization, 49, 53-54, 241
Coburn, Robert C., 82-83
Collins, Robin, 50, 177
conventionalism, *see* metric conventionalism
Conway, David, 228
Copan, Paul, 211, 212, 216
correspondence, view of truth as, 117, 125, 251
cosmic time, *see* time: cosmic
cosmology, 21, 50, 56, 62, 64-65
  quantum, 50
Craig, William Lane, 11, 12, 75, 102, 112, 113, 164, 165, 177, 212, 216, 221, 223, 224, 225, 228, 236, 237, 244, 263
creatio ex nihilo, 50, 210-215
creation, 16-20, 25, 65-66, 78, 81, 84, 86-89, 112, 180, 204, 210-215, 217-219, 229-236, 241, 243, 259, 265
Cresswell, M. J., 153, 165
C-series, 193
Currid, John D., 17
Cushing, James T., 39, 57

Dales, Richard C., 211
Davidson, H. A., 211
Davies, Paul C. W., 20-22, 24, 56, 217-218
Davis, Stephen T., 70, 91, 113
deliberation, *see* God: and deliberation
Dennett, Daniel, 79

Descartes, 207-209
differential experience of past and future, 136-139
direction of time, *see* time: direction of
Doctrine of Arbitrary Undetached Parts, 204-206
Dodd, C. H., 18
dread of future events, *see* differential experience of past and future
Dummett, Michael, 154, 165
dynamic theory of time, 12, 115-165, 169-170, 173-174, 180, 182-183, 188, 190-191, 193-194, 214

Eddington, Arthur S., 60-64, 74
Einstein, Albert, 32-33, 39-66, 69-70, 74, 75, 91, 93, 95, 167-180, 215
electrodynamics, 36-37
empty time prior to creation, *see* time: prior to creation
endurantism, 200-201, 205, 207, 209
EPR experiment, *see* Bell's Theorem
eternal present, 89-90, 108
eternity, 11-12, 13-27, 29, 31, 33, 43, 45, 47, 51, 53, 67, 77, 89-92, 94, 96, 103, 111, 112, 115, 154, 185, 211, 213, 227, 229, 231, 232, 233, 234, 235, 236, 239, 241, 243
ET-simultaneity, 90-92
Euclid, 225
extrinsic change, *see* change: extrinsic

facts
  tensed, 97-111, 115-117, 120, 122, 124, 125, 126, 129, 136-137, 156, 163, 184, 196, 199, 240, 243
  tenseless, 99, 104-112, 115-129, 136-137
Ferngren, G. B., 177
Ferré, Frederick, 198
Findlay, J. N., 53
Fine, Arthur, 178
finitude of the past, 217-233, 235
FitzGerald, G., 38, 171
Fitzgerald, Paul, 44, 55, 71, 75
flow of time, *see* time: direction of
foreknowledge, 31, 72, 243-265
Foster, Thomas R., 164
four-dimensionalism, *see* perdurantism
fragmentation of reality on, 170-171
Frankfurt, Harry, 261
Fraser, J. T., 14, 218, 219
Freddoso, Alfred J., 263
freedom, 32, 84, 86, 184, 243-244, 249, 256-263, 265
Friedman, Alexander, 60-64
Friedman, Michael, 59
Friedman, William, 130

future contingents, 243-244, 248-250, 253-256, 263-265

Gale, George, 50
Gale, Richard M., 75, 80-81, 113, 115, 144, 163, 191-192, 216
Galileo Galilei, 35-36
Gamow, George, 56, 222
Geach, Peter, 215
General Theory of Relativity, 32, 53, 57-66, 95, 168, 178-179, 187
Gilkey, Langdon, 212
Gingrich, F. W., 19
God
  and conspiracy of nature, 176
  and deliberation, 82-84
  and temporality, 77-113
  and timelessness, 29-75
  as Creator, 14, 20-22, 24, 70, 87, 96, 111, 178, 211, 212
  contingent timeless or temporal, 86-87
  eternity of, 14-27, 45, 77, 82, 219, 239
  immutability of, 29-32, 74, 83, 101, 239
  maximal cognitive excellence of, 71
  omnipotence of, 30, 249, 254
  omnipresence of, 45
  omniscience of, 11, 83-84, 96, 100, 101, 104-109, 217, 240, 243-265
  real relation to the world, 86-97, 109, 111
  simplicity of, 29-32, 73, 83, 88, 239
  temporal or atemporal, 29-113, 217
  unembodied Mind, 71
Gorenstein, M. Y., 57
Gorman, Bernard S., 115
Griffin, David Ray, 45, 158
Grünbaum, Adolf, 158, 165, 181-188, 198

Haack, Susan, 259
Harre, Rom, 159
Hasker, William, 91, 253, 255
Hawking, Stephen W., 21-22, 218-219
Healey, Richard, 49-50, 75, 195
Heller, Mark, 202-208
Heller, Michael, 57
Helm, Paul, 16-17, 27, 81, 91, 111-112, 196, 213, 237
Hendry, George S., 212
Hepburn, R. W., 73
Heraclitus, 154
Hestevold, H. Scott, 164, 201, 216
Hilbert, David, 222-225, 232
Hill, William J., 113, 252
Hoffmann, Banesh, 40, 70
Holton, Gerald, 47, 75
Horwich, Paul Gordon, 163, 165, 189
Hoy, Ronald C., 216
Hughes, Christopher, 32, 74
Hunt, David P., 261

Husserl, Edmund, 130
hyper-time, 23-26, 145-147, 150, 155, 157, 169

illusion of time, *see* temporal illusionism
Illy, Jozsef, 53, 75
incompleteness of temporal life, 67-74, 240
indexicals, 118, 127-129
inertial frame, 24, 35-49, 57-58, 62, 64, 90, 93, 96, 169-171, 174
Infeld, Leopold, 168
infinite past, *see* finitude of the past
intrinsic change, *see* change: intrinsic
Irenaeus, 211
Ives, Hubert E., 175

Jantzen, Grace M., 71
John Damascene, 85
John Duns Scotus, 77
Jubien, Michael, 202, 209

Kanitscheider, Bernulf, 53, 57, 59, 60
Kennedy, John, 54
Kerszberg, Pierre, 64, 65
Klimek, Zbigniew, 57
knowledge
  de dicto, 129
  de praesenti, 108, 118, 129
  de re, 82, 107
  de se, 82, 107-109, 129
Kretzmann, Norman, 69, 89-92, 97, 113
Kroes, Peter, 189, 215
Kvanvig, Jonathan L., 100-102, 104, 108, 113, 128

Larmor, Joseph, 38, 55
Larson, E. J., 177
Le Poidevin, Robin, 112, 149-154, 164, 215
Leftow, Brian, 14, 67, 72, 75, 81, 91, 92-97, 103-109, 112, 213, 231-232, 233
Leibniz, Gottfried Wilhelm, 66, 78, 230, 232
length contraction, 42, 173, 175, 177
Lewis, David, 129, 148, 216
Lewis, Delmas, 91, 216
Linde, Andrei, 218
local time, 61-64
Logos doctrine, 18, 241
Loizou, Andros, 159, 165
Lorentz transformation, 38
Lorentz, H. A., 38-39, 42, 53, 54-56, 64, 75, 93, 171-172, 175, 176, 239
Lorenz, Dieter, 171
Lowe, E. J., 112, 150, 164
Lucas, John R., 46, 47, 80, 234

MacBeath, Murray, 135, 137, 164
Mach, Ernst, 39, 47, 75
Mackie, J. L., 223, 225, 228-229

Mann, William E., 79, 81
Matson, Wallace, 223
Maudlin, Tim, 56, 177
Maxwell, James Clerk, 36-37, 175
May, Gerhard, 211
McCrea, W. H., 64
McGilvray, James A., 215
McGuire, J. E., 46, 74
McTaggart, J. M. E., 9, 142-154, 156, 163, 164, 165, 190, 193, 194, 195, 216
Meinhold, Arndt, 19
Mellor, D. H., 119-126, 133-137, 142, 149, 151, 153, 164, 165, 190-196, 216
Merricks, Trenton, 148, 201, 216
metaphysical relativity, 90, 96
metric conventionalism, 234
metric of time, 233-235
metrically amorphous time, *see* metric of time
Michelson, A. A., 37, 51
middle knowledge, 249
Miller, Arthur I., 40, 74, 172
Miller, D. C., 176
Milne, E. A., 53, 64
Milne-McCrea cosmology, 64
mind-dependence of becoming, 188, 197, 200, 240
Minkowski, Hermann, 54, 95, 151, 167-173, 215
Misner, Charles W., 60, 75
Morley, E. W., 37, 51
Morris, Thomas V., 32, 74, 261
Muller, R. A., 57
Munitz, Milton K., 53

Nelson, Herbert J., 81, 113, 212
Nerlich, Graham, 169, 215
New Tenseless Theory of Language, 119-129
Newton, Isaac, 25, 33-37, 44-58, 64-66, 74, 77-78, 95, 154, 175, 185, 219, 230, 232, 239, 241
Newtonian time and space, *see* absolute space, absolute time
non-metric time, *see* metric of time
Norton, John D., 75
Novikov, Igor D., 13

Oaklander, Nathan, 138, 140-141, 150, 164, 194-195
Oderberg, David S., 208
Old Tenseless Theory of Language, 117-119
omnitemporality, 29, 32, 54, 57, 63, 73, 74, 77-79, 86-87, 97, 99, 104, 109, 111, 151, 179, 199, 240
Oppy, Graham, 223

Padgett, Alan G., 15-19, 27, 113, 234, 237
Park, David, 218, 219
passage of time, *see* time: direction of

Peacocke, Arthur, 212
Penrose, Roger, 218, 219
Penzias, A. A., 56
perdurantism, 201-209, 215
Perry, John, 118-119, 129, 164, 183
personal identity, 205-206
personhood and timelessness, 79-86, 240
Philo of Alexandria, 18
Philoponus, John, 211
physical time, *see* time: measured
Pike, Nelson, 81, 212
Plantinga, Alvin, 123-124, 131
Plöger, Otto, 18-19
Podlaha, M. F., 171
Podolsky, Boris, 54-55
Poincaré, Henri, 51-53, 55, 172
Polkinghorne, John, 211-212
Popper, Karl, 56, 57
positivism, *see* verificationism
presentism, 102, 148-154, 156
presentness of experience, 120, 134-137, 143, 163, 199
Price, Huw, 132, 163, 165, 184, 189
Principle of Relativity, 58-59, 168
Prior, Arthur N., 99, 136-137, 148, 156, 164, 165
privileged reference frame, 173-174, 176
Prokhovnik, Simon J., 178, 215

quantum cosmology, *see* cosmology: quantum
quantum mechanics, 50
Quinn, Philip L., 75, 233

Recanati, François, 164
Rees, Martin J., 62-63
real relation of God to creatures, *see* God: real relation to the world
reference frame, 35, 41, 54, 56-57, 59, 74, 90-92, 96-97, 103, 170-171, 176
relationalism, *see* time: relational view of
Relativity Theory, *see* General Theory of Relativity, Special Theory of Relativity
relief over past events, *see* differential experience of past and future
Rescher, Nicholas, 177, 251-252
Robertson, H. P., 61
Rosen, Charles, 73
Rosen, Nathan, 55
Ross, Hugh, 24-26
Rovelli, Carlo, 186-187
Rucker, Rudolf v. B., 224
Ruderfer, Martin, 175
Rudnicki, Konrad, 57
Russell, Bertrand, 117, 127, 188

Schleiermacher, F. D. E., 212
Schlesinger, George N., 138-140, 164
Schmidt, Johannes, 17

Schücking, E. L., 64
self-ascription of properties, 129
Sellars, Wilfrid, 12
Senor, Thomas, 237
sensorium, 47
sentence tokens, 119-122, 125, 250
sentence types, 125-126
Shoemaker, Sydney, 78
simplicity of God, *see* God: simplicity of
simultaneity, 32, 40, 42, 46, 48-49, 52-53, 56, 59, 62, 89-92, 96, 103, 169, 171, 173, 176
  absolute, 46, 52-57, 62, 74, 151, 173, 176
singularity as a boundary to time, 21, 217-218, 236
Sklar, Lawrence, 49, 51, 75, 139, 162, 165, 187
Smart, J. J. C., 130, 155, 156-157
Smith, Quentin, 14, 113, 115, 164, 221, 225, 237
Smoot, G. F., 57
solipsism, 151, 153
Sorabji, Richard, 18, 157, 211, 224, 228, 237
sorites paradoxes, 207-208
spacetime, 217-218
Special Theory of Relativity, 12, 23, 32, 38, 40, 43-44, 46, 47, 49, 51, 54-59, 62-63, 66, 74, 90-91, 95-96, 113, 151, 167-180, 186, 188
Squires, Euan, 50
static theory of time, 12, 70, 110-112, 137-140, 142, 145, 148-150, 158, 162, 163, 167-216
string theory, 179
Stump, Eleonore, 69, 89-92, 97, 234, 237
Suppe, Frederick, 49
Swinburne, Richard G., 220, 226, 234, 237

Talbott, Thomas Bradley, 252
Taliaferro, Charles, 75, 233
Tarski, Alfred, 121
Tatian, 211
Taylor, Edwin F., 74, 172, 215
Taylor, J. G., 74, 172, 215
temporal becoming, 22, 70, 95, 97, 109-111, 137-144, 148, 154-161, 169, 173-174, 180-184, 192-193, 197-200, 215, 217, 229, 235, 239-241
temporal illusionism, 199
temporal indexicals, 111, 127
temporal necessity, 259-263
temporal parts, 200-210, 214
temporal relations, 96, 108, 111, 115, 150, 152, 162, 188-197
temporality, divine, *see* omnitemporality
tensed facts, *see* facts: tensed
tensed theory of time, *see* dynamic theory of time
tenseless facts, *see* facts: tenseless

tenseless theory of time, *see* static theory of time
Theophilus, 211
Thomas Aquinas, 29-30, 31, 71, 77, 88-89, 97, 113, 152
Thomsen, Judith Jarvis, 203
Thorne, Kip S., 60, 75
three-dimensionalism, *see* endurantism
time
    absolute, 25, 33-34, 38-39, 43-54, 63-66, 74, 77, 151, 154, 162, 174, 231
    asymmetry of, 138, 160-163
    beginning of, 17-20, 211, 217-219, 233
    cosmic, 23, 53, 60-66, 95, 214
    direction of, 23, 67, 70, 141, 154-163, 179, 186
    dynamic vs. static, *see* dynamic theory of time, static theory of time
    measured, 13, 46, 182-187, 219, 236
    mental events sufficient for, 66, 94, 96, 180, 219
    metaphysical vs. measured, 46-47
    metric of, *see* metric of time
    metrically amorphous, *see* metric of time
    prior to creation, 66, 219, 230-235
    reductionistic views of, 185
    relational view of, 66, 83, 86, 93, 230
    tensed vs. tenseless, *see* dynamic theory of time, static theory of time
time dilation, 42, 173-178
time travel, 263
time's flow, *see* time: direction of
Tipler, Frank J., 218
Tooley, Michael, 115, 190-192
Trinity, 26, 30, 85, 109, 241
truth bearers, 125
truth makers, 122-126
Twin Paradox, 172

universe, beginning of, *see* beginning of the universe

van Inwagen, Peter, 123, 154, 156, 204-209, 216

verificationism, 39, 47-52, 66, 74
Vilenkin, Alexander, 218

Walker, A. G., 61
Walker, Ralph C. S., 112
Waterlow, Sarah, 161, 165
Weinberg, Steven, 179
Wells, H. G., 70
Wessman, Alden E., 115
Westermann, Claus, 17, 211, 216
Wheeler, John Archibald, 13, 55, 60, 75, 172, 215
Whitehead, Alfred North, 157-158
Whitrow, G. J., 27, 59, 62-63, 95, 165, 180, 215
Why did not God create the world sooner? 229-233
Why is it now? 182-183
Whybray, R. N., 18
Wierenga, Edward R., 101-105, 113
Wilder, Laura Ingalls, 68
Williams, Donald C., 155, 165
Wilson, R. W., 56
Winnie, John A., 171
Wisdom, John, 164
Witherington, Ben, III, 248
Wittgenstein, Ludwig, 225-226
Wolfson, H. A., 211
Wolterstorff, Nicholas, 131, 165, 237
world map, 23, 53

Yancey, Philip, 22-23
Yates, John C., 81, 91, 112, 213

Zahar, Elie, 75
Zeno, 143, 158, 181, 188

Scripture Index

Genesis
1:1   17, 18, 20, 210, 211
1:5   17

Numbers
23:19   248

Deuteronomy
18:22   245

1 Samuel
15:29   248

Psalms
8:3-8   178
33:9   211
54:20   19
55:19   19
90:2   15, 211, 213
104:5-9 211
106:48   11
139:1-6 247
139:7-10 16

Proverbs
8:22-23 17, 18
8:23   19
8:27-29 211

Isaiah
38:1-5   248
40:21   18
41:4   14
41:4, 26 18
41:21-24 245-246
44:24   211
45:18, 24 211
46:9-10 244
46:10   210
57:15   14

Jeremiah
23:23-24 16
26:3   247
36:3   247

Amos
7:1-6   248

Jonah
3   248

Malachi
3:6   31

Matthew
13:35   213
24   245
24:21   213
25:34   213

Mark
13   245

Luke
11:50   213
21   245

John
1:1-3   17, 211
17:24   20, 213

Acts
11:27-28 245
13:1   245
15:32   245
21:9   245
21:10-11 245

Romans
4:17   211
11:36   211
16:25   19

1 Corinthians
2:7   19
8:6   211
12:28-29 245
14:29, 37 245

15:54   214

Ephesians
1:4   20, 213
1:10   244
3:9   244
3:11   244
4:11   245

Colossians
1:16   211

2 Timothy
1:9   19
1:9-10   244

Titus
1:2-3   19

Hebrews
1:2-3   211
1:3   212
1:10-12   14
9:26   213
11:3   211

James
1:17   31

1 Peter
1:20   20, 213, 245

Jude
25   19, 213, 241

Revelation
4:8b   15
4:11   211
13:8   20, 213
17:8   213
22:6   245

EXTRA-BIBLICAL LITERATURE

1 QS
3:15   211

2 Baruch
21:4   211

2 Enoch
25:1ff   211
26:1   211
65:6-7   16, 19

2 Maccabees
7:28   211

Joseph and Aseneth
12:1-3   211

Odes of Solomon
16:18-19   211

Sirach
16:26   19
23:20   19
24:9   19